ACTIVE PACKAGING *for* FOOD APPLICATIONS

ACTIVE PACKAGING *for* FOOD APPLICATIONS

Aaron L. Brody
Consultant, formerly CEO
Rubbright • Brody, Inc., Duluth, GA

Eugene R. Strupinsky
Formerly, Vice President
Research and Market Development
for Plastic and Packaging
Rubbright • Brody, Inc.

Lauri R. Kline
Project Leader
Natick Soldier Center

CRC Press
Taylor & Francis Group
Boca Raton London New York

CRC Press is an imprint of the
Taylor & Francis Group, an **informa** business

CRC Press
Taylor & Francis Group
6000 Broken Sound Parkway NW, Suite 300
Boca Raton, FL 33487-2742

First issued in paperback 2019

© 2001 by Taylor & Francis Group, LLC
CRC Press is an imprint of Taylor & Francis Group, an Informa business

No claim to original U.S. Government works

ISBN-13: 978-1-58716-045-5 (hbk)
ISBN-13: 978-0-367-39728-9 (pbk)

Library of Congress Cataloging-in-Publication Data

Main entry under title:
  Active Packaging for Food Applications

Full Catalog record is available from the Library of Congress

Library of Congress Card Number 200108929

Visit the Taylor & Francis Web site at
http://www.taylorandfrancis.com

and the CRC Press Web site at
http://www.crcpress.com

Table of Contents

Preface

THIS text was derived from a draft technical report for the U.S. Army Soldier and Biological Chemical Command (SBCCOM), Natick Soldier Center (formerly the U.S. Army Natick Research, Development and Engineering Center), prepared by Rubbright•Brody, Inc., entitled "Active Packaging Applications to Enhance the Preservation and Quality of Contained Foods." The purpose of this original report was to identify and characterize domestic and international active packaging technologies with potential for direct U.S. military application. The report reviews commercial and developmental technologies concerning oxygen-scavenging, odor and moisture removal, aroma emission, and antimicrobial systems to better retain the quality of contained food.

With the gracious cooperation of those who generated the original document and the permission of all of the sponsoring parties, this book updates, expands, and enhances the original text to the broader focus of the commercial world.

This book focuses on active packaging in all of its manifestations. As we indicate in this text, literally thousands of citations from the peer-reviewed, trade, commercial, and patent literature were found, classified, and reviewed. Those who wrote the references and those who did not but who worked on the laboratory benches, in the pilot plants, and on the production floors were interviewed in depth and became our contributing authors and critics.

We have highlighted oxygen-scavenging packaging because it has the longest history and the largest use. More significantly, oxygen control is a justifiably growing technology that will ultimately graduate from adjunct to full status as a means to enhance food and beverage preservation.

The role of oxygen-scavengers in this evolving food-preservation enhancement has not yet been defined but is almost certain to become significantly more important in the future. The same is true for the other active packaging technologies discussed in this book: odor control, moisture control, carbon dioxide control, and antimicrobials.

Carbon dioxide is no longer merely a gas that imparts a tongue tingle sensation to beverages, but rather an integral component of much modified-atmosphere packaging in which the physiological biochemistry is altered and manipulated to prolong quality retention. Colors, flavors, and pathways may be disrupted. Optimal concentrations in the dynamic environment of living tissue are changed by the very presence of carbon dioxide as quality retention is maximized.

Thus, like oxygen control, the independent control of carbon dioxide is crucial to the preservation of post-harvest plants and of animal tissue whose microbiological populations may be suppressed.

Another challenge is the elusive target of destroying or retarding microbiological growth by means of antimicrobial agents. Among the rules of thumb with which we must contend are the following: chemicals that can control microorganisms are often toxic to humans; many chemicals may be selectively antimicrobial; contact of the chemical with the microorganism is usually required; and too many of these agents impart disagreeable flavors.

Most of the numerous reports extolling the virtues of antimicrobial agents in package materials minimize the caveat that direct contact is required. What of the irregular surfaces of most foods? And what about all of the microorganisms buried beneath the surfaces? Some proposed agents, such as ethyl alcohol or allylisothiocyanate, can be vaporized to permeate foods and thus extend their effectiveness. But what about their residual flavor and selectivity? And regulatory implications? These and many other issues are raised and explored in this meticulously researched volume.

Can package materials and structures be truly active—sensing and reacting in a positive manner while remaining benign to other properties? This book will suggest that technology has progressed far in offering to control oxygen and, to some degree, in controlling moisture, odor, and microorganisms. As of this writing, however, we have not yet arrived at a point where we can depend on a package material or structure. We must continue to think and act holistically—to combine the tried, such as barrier and clean operations, with our new technologies. Then we should consider hurdle technologies—combining oxygen control with antimicrobials, for example.

Some have learned that aseptic packaging may become hurdle technology for extended shelf life refrigerated preservation. Now, if we stretch our thinking, we can embrace our as-yet-limited knowledge of active packaging to enhance quality in concert with existing technologies.

If this book teaches anything, it is that active packaging is still in its elementary stages. If we can learn something, it is that we should succumb to the wiles of how easy it might be to read a paragraph out of context and instantly cry out, "This is the answer!" Like its ancestors of gas-barrier, retort-pouch, and modified-atmosphere packaging, active packaging has a distance to travel

Together with its acknowledged antecedent, Michael Rooney's *Active Food Packaging*, this book should help propel us along the path of appreciating and understanding the challenges and opportunities that we shall surely gain from active packaging.

Acknowledgements

The authors wish to gratefully acknowledge the special contributions of Mr. Michael Rooney, who drafted part of the original section on oxygen-scavenging; Dr. Jung Han, who contributed a chapter on antimicrobial packaging; and Mrs. Lauren Milch, Physical Scientist, and Dr. William Porter, Chemical Technical Consultant, of the U.S. Army Natick Soldier Center, Natick, Massachusetts.

We received remarkable cooperation from the community of researchers who have been intimately involved with active packaging for the past decade or more: Mr. Michael Rooney of Food Science Australia, formerly Australia's Commonwealth Scientific Industrial and Research Organization (CSIRO); Dr. Joseph Hotchkiss of Cornell University; Mr. Ron Idol, formerly of Multisorb Technologies, Inc., and now with Air Liquide America; Dr. John Floros, formerly of Purdue University and now Chair, Department of Food Science and Technology, Pennsylvania State University; Dr. John Krochta of University of California, Davis, and his student, Dr. Jung Han, now Assistant Professor, University of Manitoba, Canada; Dr. James P. Smith of McGill University, Canada; and Dr. Boh C. Tsai, formerly of Amoco Chemicals and now with George O. Schroeder Associates. Each offered considerable information that was employed extensively throughout the preparation of this manuscript.

Much appreciation is expressed to Wendy S. Hill for her marvelous manuscript preparation, editing, and watchful care over the many iterations. Her conscientious supervision and dedication to the precision of this task were among the indispensable foundations of the quality to which we have striven.

Co-author Strupinsky wishes to thank his wonderful wife, Claire, who tolerates much more than anyone has a right to expect.

Co-author Kline is grateful for the loving support of her husband, Gary; her family, Linda, Harold, David, and Marc Pruskin; and of course, Lucy. She thanks her many talented and dedicated colleagues at Natick Labs.

Co-author Brody wants this book to be a visible tribute to his wonderful family to which he is forever devoted and whom he loves: grandchildren Pierce Aaron and Natalia Siena and their parents, Robyn Todd and Kellie; grandchildren Skyler Alexis and Camryn Alexander and their parents, Glen Alan and Sharon; grandchildren Derek Jason and Michelle Jennifer and their parents, Stephen Russell and Susan; and Carolyn, beloved wife, mom, and now Grandma to our entire family.

Abbreviations, Acronyms, and Symbols

Å	Angstrom unit
ABS	acrylonitrile butadiene styrene
Ag	silver
Ag^+	silver ion
AgCu	silver copper
Ag_2O	silver oxide
AgZn	silver zinc
AIT	allylisothiocyanate
Al^{+++}	aluminum ion
ASTM	American Society for Testing Materials
atm	atmosphere pressure
Au^{++}	divalent gold ion
a_w	water activity
Ba^{++}	barium ion
BHA	butylated hydroxy anisole (antioxidant)
BHT	butylated hydroxy toluene (antioxidant)
Ca^{++}	calcium ion
$Ca(OH)_2$	calcium hydroxide
CAP	controlled atmosphere packaging
cc	cubic centimeter
Cd^{++}	cadmium ion
cfu	colony forming unit
ClO_2	chlorine dioxide
cm^3	cubic centimeter
CMC	carboxymethyl cellulose
Co	cobalt
Co^{++}	cobaltous ion

CO_2	carbon dioxide
CPET	crystallized polyester
CSIRO	Commonwealth Scientific and Industrial Research Organization (Australia)
Cu^{++}	cupric ion
dm^2	square decimeter
E. coli	*Escherichia coli* (microorganism)
e.g.	for example
EDTA	ethylenediaminetetraacetic acid (chelating agent)
EPA	Environmental Protection Agency
EVA	ethylene vinyl acetate
EVOH	ethylene vinyl alcohol copolymer
FDA	Food and Drug Administration
Fe	iron
Fe^{++}	ferrous iron/ferrous cation
Fe^{+++}	ferric ion
Fe_2O_3	iron oxide/ferric oxide
$FeSn_6$	tin/iron — tin alloy
$FeSO_3$	iron sulfite/ferrous sulfite
$ft.^2$	square foot
g	gram
GRAS	generally recognized as safe (by FDA)
H or H_2	hydrogen
H_2O	water
H_2O_2	hydrogen peroxide
H_2S	hydrogen sulfide
H_2SO_4	sulfuric acid
HDPE	high density polyethylene
Hg	mercury
HPMC	hydroxypropyl methyl cellulose
i.e.	that is
$in.^2$	square inch
IR	infrared
J	joule
K^+	potassium ion
kg	kilogram
$KMnO_4$	potassium permanganate
kW	kilowatt
LDPE	low density polyethylene
Li^+	lithium ion
LLDPE	linear low density polyethylene
M	mole
m^2	square meter

MA	modified atmosphere
MAP	modified atmosphere packaging
MC	methyl cellulose
MA	methacrylic acid
mEq	milliequivalent
mg	milligram
Mg^{++}	magnesium ion
MIC	minimum inhibitory concentration
min	minute
ml	milliliter
mm	millimeter
Mn	manganese
Mn^{++}	manganous ion
MRE	Meal, Ready-to-Eat
Na^+	sodium ion
NaCl	sodium chloride
NH_3^+	ammonium ion
Ni^{++}	nickelous ion
O_2	oxygen
O_2TR	oxygen transmission rate
OPET	oriented polyester (film)
OPP	oriented polypropylene (film)
OS	oxygen scavenging (Cryovac Sealed Air designation)
OTC	odor and taste control (DuPont designation)
P	permeation coefficient
PAA	polyalkyl acrylate
PAI	polyalkylamines
Pb^{++}	lead (plumbous) ion
Pd	palladium
Pd^{++}	palladium ion or palladous ion
PDA	potato dextrose agar
PE	polyethylene
PEI	polyethylene imine
PET	polyethylene terephthalate (polyester) (common tradenames of the film version are Melinex and Mylar, both DuPont)
pH	log of the reciprocal of hydrogen ion concentration (pH 7 = neutral, pH 1 = acid, pH 14 = basic)
PP	polypropylene
ppb	parts per billion
ppm	parts per million
PS	polystyrene
psig	pounds per square inch (gauge)
PVA	polyvinyl alcohol

PVdC	polyvinylidene chloride
PVdC	copolymer saran (copolymer of polyvinyl chloride and polyvinylidene chloride)
PVOH	polyvinyl alcohol
RH	relative humidity
SBCCOM	U. S. Army Soldier and Biological Chemical Command (Natick, Massachusetts, USA)
SCC	side chain crystallizable
SiO_2	silicon dioxide
Sn	tin
Sn^{++}	stannous ion
Sn^{++++}	stannic ion
SO_3^{--}	sulfite ion
SO_4^{--}	sulfate ion
SPI	soy protein isolate
Sr^{++}	strontium ion
TAPPI	Technical Association of the Pulp and Paper Industries
TBHQ	terbutylhydroquinone (antioxidant)
TBTO	tributyl tin oxide
UGR	Unitized Group Ration
ULDPE	ultra low density polyethylene
UV	ultraviolet
Zn^{++}	zinc ion
°C	degrees Celsius/Centigrade
°F	degrees Fahrenheit
μl	microliter
μm	micrometer
μmoles	micromole
Zn	zinc

CHAPTER 1

Introduction

A CTIVE packaging, sometimes referred to as interactive or "smart" packaging, is intended to sense internal or external environmental change and to respond by changing its own properties or attributes and hence the internal package environment. Active packaging has been considered a component of the packaging discipline for several decades or since the first inclusion of desiccants in dry product packages. In their own moisture-permeable sachets, desiccants absorb water vapor from the contained product and from the package headspace, and absorb any water vapor that enters by permeation or transmission through the package structure. As separate entities within packages, active packaging sachets, pouches, patches, coupons, labels, etc., are not often integral to the package—a semantic differentiation.

Desiccant pouches are widely used in the packaging of hardware and metal goods. The best-known and most widely used active packaging technologies for foods today are those engineered to remove oxygen from the interior package environment. Oxygen scavengers reduce oxidative effects in the contained product. Most oxygen scavengers in commercial use today are gas-permeable, flexible sachets containing reduced iron (i.e., iron not in the fully oxidized state) particles inserted into food and other packages from which air is initially removed by vacuum or by flushing with inert gas During the last two decades of the twentieth century, commercial incorporation of oxygen-removal materials directly into a package structure occurred with varying results. Several applications for beer and juice bottles became commercial in 2000.

The goal of active packaging, in conjunction with other food processing and packaging, is to enhance preservation of contained food and beverage products. For example, to optimize the effects of oxygen scavenging, oxygen should first be removed from the product during processing and packaging operations. The oxygen must also be thoroughly removed from the package

1

interior and the package materials, and the package structure, including materials and closure, must be barriers to further oxygen entry. In other words, oxygen scavenging complements good oxygen-control practices.

In addition, oxygen is certainly not the only vector that can influence the quality of the contained food. For example, moisture gain or loss, light, non-oxidative reactions, microbiological growth, and enzymatic activity may all, individually or collectively, be involved in food-product deterioration.

Worldwide development efforts devoted to oxygen removal have indicated that analogous efforts by the same and parallel research teams continue to be applied to oxygen scavenging and are being studied for other active packaging forms. Michael Rooney's book, *Active Food Packaging* (Rooney, 1995), coalesced some of the many known active packaging concepts for foods into a single volume. This book did not, however, probe deeply into some of the more promising technologies that have been proposed and have entered the marketplace, including antimicrobial films, carbon dioxide emitters, aroma emitters, and odor absorbers. Each of these is discussed or referred to in this text. On review of the commercial situation in the United States, and especially the application of oxygen-scavenging compounds into the walls of beer bottles and processed-meat packages, the reasons for the notable paucities in the Rooney book become apparent. Although reviewed and referenced in commercial contexts, definitive scientific documentation and publications are lacking, unsubstantiated, or unclear. Descriptions of plastic bottle wall oxygen scavenging appear only in the patent literature, which does not, of course, detail specifics. The 1999 and 2000 George O. Schroeder conferences, "Oxygen Absorbers: 2000 and Beyond" and "Oxygen Absorbers: 2001 and Beyond" (Anonymous, 1999, 2000) set out to probe the expanding realm of oxygen-removal mechanisms. The presentations offered excellent reviews of historical and contemporary technologies and scientific studies but did not elucidate on the intriguing, but still largely proprietary, industrial world of oxygen scavenging.

Nevertheless, many of the active packaging systems not fully discussed in the book *Active Food Packaging* have reached the commercial market, some with notable success. High-gas-permeability films, including some that increase their oxygen permeability with increasing temperature, are used for packaging fresh-cut produce. Use of these temperature-sensitive package materials is expected to increase because the technology developer has acquired a fresh-produce packager who, of course, uses the technology in its package materials.

Carbon dioxide and ethylene scavengers for modified-atmosphere (MA) or, more precisely, controlled-atmosphere (CA) food preservation are common in large bulk shipments. Carbon dioxide emitters to suppress microbiological growth have experienced limited success in modified-atmosphere packaging (MAP). Ethylene scavengers are among the more successful commercial active packaging technologies in the fresh-fruit bulk-shipment category.

Odors generated or captured within closed food packages are undesirable, and their obviation has been a research topic for years. Odor removers incorporated into packaging are increasingly important in some classes of food packaging.

Antioxidants and oxygen interceptors incorporated into package materials, such as tocopherols (vitamin E), have emerged in recent years and are increasingly employed to combat odors generated in plastic processing. Tocopherols, which are nonvolatile, have not replaced volatile butylated hydroxyanisole/butylated hydroxytoluene (BHA/BHT) which migrate into foods in product antioxidant applications, but they appear to be new antioxidants of choice for mitigating the effects of oxygen. Entities such as oxygen scavengers/interceptors react with oxygen to form new compounds. Oxygen absorbers may remove oxygen by any means, including physical. Antioxidants react with free radicals and peroxides to retard or block the actual oxidation reactions. Sequestering agents tie up inorganic catalysts that might otherwise accelerate adverse oxidative reactions.

Members of the food technology and packaging communities have long regarded package materials as an ideal reservoir and delivery vehicle for antimicrobial compounds. For many years, sorbic acid has been incorporated sparingly on the interior of package structures as an antimycotic in a limited number of dry food packages. The obvious benefits of sorbic acid as a mold and yeast inhibitor have been one foundation by which numerous other antimicrobial agents have found their way into food package materials. Unfortunately, most antimicrobial agents also exhibit toxicity when they enter the food from the package and would be consumed as part of the food. Thus, actual commercialization has been proceeding slowly, except in Japan where several compounds have been reported to function effectively as antimicrobials in commercial packages.

As with oxygen scavengers, the major technological and commercial successes for antimicrobials have been achieved by Japanese organizations for packaging Japanese products in Japan. Nevertheless, the concept of integrating microbistatic and microbicidal materials and plastic packaging has been very attractive. Numerous attempts have been and are being made to translate favorable laboratory results into safe and effective commercial food packaging. The growing list of successes in active packaging beyond oxygen scavenging has been noted by the food-packaging community.

OXIDATIVE DETERIORATIVE REACTIONS

Despite the best of barriers, processing technologies, and controls, foods are susceptible to biochemical and other forms of deterioration. For example, all thermally stabilized foods in hermetically sealed packages under ambient-temperature conditions undergo both oxidative and nonoxidative degradation during storage and distribution. Residual oxygen within both the food and headspace of hermetically sealed barrier containers reacts to oxidize food components. Simultaneously, nonoxidative reactions involving appearance, mouth feel, and flavor occur, with all of the reactions proceeding at more rapid rates at higher temperatures. With few exceptions, such as some wines and cheeses, the longer the time in distribution, the lower the sensory quality of the packaged food.

At the extreme—even in commercially sterile packages—thermophilic microorganisms, of relatively little consequence in normal commercial distribution but of great importance if ambient temperatures of distribution are in the 49°C (120°F) and above range (e.g., Middle Eastern desert conditions), can grow and spoil the food or even render it hazardous if they are facultative anaerobes. Enhancing the package oxygen barrier or removing residual oxygen from the product will have relatively little value in these situations. Thus, today oxygen control is aimed at better retention of biochemical quality. Nevertheless, active packaging has been examined as a possible source of remediation for deterioration in very long or extreme distribution conditions.

Except for the extreme cases noted above, contemporary commercial food products move through distribution channels faster and with better temperature control than in the past.

ODOR CONTROL

In-package odor removers have the potential to scavenge the malodorous constituents, i.e., stinks, of both oxidative and nonoxidative biochemical deterioration. Aroma, fragrance, and scent emitters have been suggested to impart odors that can mask unpleasant odors, react with the undesirable odors to neutralize them, or provide desirable "fresh-like" smells. Certainly, the not-uncommon practice of incorporating fragrances into plastic film structures and materials for garbage and trash bin liners and for containers and cosmetic samples suggests a potential for aroma-emitting packaging. On opening a food package, whether after a very long distribution time and/or at above-ambient temperature conditions or after only a short distribution period at below-ambient temperature conditions, the emission by the package structure/material of a desirable aroma would represent a food quality enhancement. This effect would be not unlike incorporating pleasant aromas into the headspaces of packages of food such as instant coffee or juice concentrate to deliver a pleasant and enticing olfactory effect to the consumer on initial opening. Many food products generate low levels of natural undesirable deteriorative odors, often called "confinement odors," which cannot be overcome by adding back flavor—but might be mitigated by the use of aroma-emitting package structures.

MINIMALLY PROCESSED FOODS

In another important contemporary market, that of minimally processed prepared foods, the foods are processed and packaged with relatively little or no heat to reduce thermal damage and thus deliver better quality while prolonging the preservation time, usually under refrigerated conditions. A major issue with such foods is that they are not sterile and so reduced oxygen is often required to achieve reasonable commercial shelf life from both microbiological safety and biochemical deterioration perspectives. Due to the presence of microorganisms, the risk of microbiological hazard is increased over more conventional short-term-distribution foods, which are preserved by pH control or chemical additives. Among the mechanisms suggested to obviate the microbiological safety and spoilage potential in minimally processed foods is the incorporation of antimicrobials into the package materials.

Most antimicrobials employed in Japan and tested in the United States generally function only by contact between the antimicrobial in the package material and the food, which is a somewhat limiting factor. Contact microbicidal or microbistatic action can be desirable. In recent years, developments have been underway in the area of "natural" and "controlled-release" antimicrobials. Conceptually, both have potential for retarding food spoilage, but they have not been fully evaluated. Another explored concept is that of antimicrobials that volatilize from the package materials and function without direct contact between package materials and the food surface. Such agents should have the additional ability to permeate through and beneath the surfaces of the food and reduce any microbiological threat.

ACTIVE PACKAGING OPPORTUNITIES

The range of active packaging is so broad that, with further development, many of these technologies will be able to aid in the preservation and quality retention of commercially processed and packaged food (Tables 1, 2, and 3). The challenge is that the numbers of different active packaging proposals and commercializations from around the world are very large. Further, many of these claims are often difficult to comprehend. Perhaps in the future, some of these active packaging concepts may be further developed into systems that are or would be applicable in the United States and Europe.

The research activity culminating in this work on active packaging has been challenging as a result of the continuing input of products and concepts emerging from each phase. The extensive bibliography, which has been abbreviated to render it more useful, points out the thousands of attempts in various aspects of active packaging.

6   INTRODUCTION

TABLE 1. Types of Active Packaging Systems with Mode of Action and Representative Manufacturers (Excluding Oxygen Scavengers).

System/Action	Substance	Organizational Source
Ethylene absorbing	• Activated carbon/potassium permanganate	• Kuraray/Nippon (Japan) • Greener (Japan)
Ethanol emitting	• Micro-encapsulated ethanol	• Freund (Japan)
Moisture absorbing	• Polyvinyl alcohol encapsulation	• Grace Chemical (Davison)
	• Silica gel	• Capitol Specialty Plastics • Multisorb Technologies
	• Clay-based	• Sud-Chemie Performance Packaging
Antimicrobial releasing	• Sorbates • Benzoates • Propionates • Silver salts • Sulfur and mercurial compounds • Bacteriocins • Sub-micrometer cell wall penetrants	• Mitsubishi Gas Chemical (Japan) • Microban Products • Various from Japan
	• Zeolites	• Mitsubishi Gas Chemical (Japan) • Shinagawa Fuel (Japan) • Tachyon Energy (Japan)
	• Chlorine dioxide	• Bernàrd Technologies • Englehard Corp
Antioxidant releasing	• BHA/BHT • TBHQ • Vitamin C or E	• Roche
Flavor/odor absorbing	• Activated carbon • Sodium bicarbonate	• Arm & Hammer • Cabot
Chemical stabilizers	• Tocopherol or vitamin E	• Roche

Adapted from Floros et al , 1997

OUTLINE

This text is divided into sections covering the major categories of active packaging based on importance and amount of research and development effort invested and commercial interest:

• oxygen scavenging
• moisture control

TABLE 2. Selected Patents on Various Active Packaging
Technologies: CO_2 Absorbers/Emitters, Ethylene
Absorbers, and Ethanol Generators
(Excluding Oxygen Scavengers).

Company	Function and Substance(s)	Patent Year	Patent Number
Freund Industrial Co Ltd (Japan)	Ethanol-vapor generator: several different substances mentioned (1 O_2 scavenger)	1989	US 4820442
J Velasco Perez	Ethylene absorber/CO_2 generator: sepiolite and $KMnO_4$	1990	US 4906398
K K Nasa (Japan)	Ethylene absorber: far-IR radiating ceramic granules	1990	US 4927651
Kyoei Co Ltd (Japan)	Ethylene absorber: zeolite (for apples) (1 O_2 scavenger)	1988	US 4759935
Mitsubishi Gas Chemical Co (Japan)	Ethanol emitter: for example, activated carbon, SiO_2, clay, celite, zeolite, paper cotton 1 acetaldehyde remover (1 O_2 absorber)	1992	EP 0505726A1
Mitsubishi Gas Chemical Co (Japan)	CO_2 generator/O_2 scavenger	1988	US 4762722
Mitsubishi Gas Chemical Co (Japan)	CO_2 absorber/O_2 scavenger; $Ca(OH)_2$	1982	US 4366179
Toppan Printing Co Ltd (Japan)	CO_2 generator/O_2 scavenger: Mn-salt 1 metal 1 alkali 1 sulfite	1983	US 4384972
Toppan Printing Co Ltd (Japan)	Ethylene absorber zeolite 1 bentonite 1 active carbon	1982	US 4337276

TABLE 3. Current and Potential Future Applications of Active Packaging Technologies (Excluding Oxygen Scavengers).

Applications	Food Groups					
	Dry	Minimally Processed	Meat and Dairy	Frozen Foods	Bakery	Beverages
Ethylene emitter		Fruit/vegetables				
Ethylene scavenger		Fruit/vegetables				
Moisture absorber	All dry foods	Fruit/vegetables, meats, etc.				
Moisture regulator						
Ethanol emitter	Semi-dry fish, meat	Prepared foods	Cheese		Sweet baked goods, bread	
Antimicrobial-releasing film		Fruit	Cheese, meat		Bread, cakes	
Antioxidant-releasing film	Breakfast cereal				Hard baked goods	Bag-in-box wine; wine
Flavor-containing and -emitting film	Cereals	Prepared foods		Ice cream		Orange juice
Color-containing film			Surimi			
Anti-stick film			Cheese slices		Frosting, candy	
Enzyme inhibitor		Fruit/vegetables				
CO$_2$ regulator		Fruit/vegetables				
Light control	Snacks, lipids	Fruit/vegetables, meats				

- ethylene removal from fresh-food packaging
- antimicrobials
- odor removal
- aroma emission

BIBLIOGRAPHY

Anonymous. 1992. "Active Packaging—On Its Way." *Australian Packaging,* 40(6): 12–14, June.

Anonymous. 1999. "Oxygen Absorbers: 2000 and Beyond." Conference Proceedings George O. Schroeder Associates, Inc Appleton, Wisconsin, June.

Anonymous. 2000. "Oxygen Absorbers: 2001 and Beyond." Conference Proceedings. George O. Schroeder Associates, Inc. Appleton, Wisconsin, June.

Downes, T. W 1995. "Films as Carriers of Functional Ingredients and Interactive Components." Institute of Food Technologists Annual Meeting.

Floros, John D., Lise L. Dock, and Jung H Han. 1997. "Active Packaging Technologies and Applications " *Food, Cosmetics and Drug Packaging,* 20(1).

Labuza, T. P. and W. M. Breene. 1989 "Applications of 'Active Packaging' for Improvements of Shelf-Life and Nutritional Quality of Fresh and Extended Shelf-Life Foods." *Journal of Processing and Preservation,* 13, pp. 1–69.

Miltz, J., P. Hoojjat, J. K. Ham, J. R. Giacin, B. R. Harte, and I J. Gray 1988 "Food Packaging Interactions." J. H. Hotchkiss, ed , *American Chemical Society Symposium Series,* 365: 33

Rooney, Michael L. (ed). 1995. *Active Food Packaging.* Glasgow, UK: Blackie Academic & Professional.

Wygonik, Mark. 1993. "Just Around the Corner." *Canadian Packaging,* 46·4, 10

CHAPTER 2

Oxygen Scavengers

INTRODUCTION

P ROTECTION of packaged food contents for extended periods under any prolonged chilled or ambient-temperature conditions requires a series of interrelated actions:

(1) The product itself must be processed in a manner that stabilizes it against enzymatic, microbiological, and biochemical activity, with the former two of concern primarily in foods packaged for short-term, chilled distribution.
(2) The product must be protected against recontamination by active biological vectors of deterioration such as viable enzymes or microorganisms, usually by hermetic sealing.
(3) The package generally should be a high gas-barrier structure that is hermetically closed, i.e., the seal is not a route for gas transmission into the package or for recontamination.
(4) The environment within the hermetically sealed package should be made free of oxygen, which is not always an easy task since most foods contain occluded and dissolved air; finite quantities of reactable oxygen are usually present in the package headspace or product interstices whether or not a headspace is present, and the package material itself might contain oxygen that can be released into the food.

Recognizing that residual oxygen can react biochemically with contained food and cause long-term adverse oxidative effects, which increase as the temperature rises, some barrier and vacuum packaging has been supplemented recently by an oxygen-scavenger component of active packaging.

Beyond moisture removal, the most important active packaging objective has been removal of oxygen using techniques variously called absorption, interception, or scavenging.

DEFINITIONS

The terms *antioxidants*, *interceptors*, *absorbers*, and *scavengers* have been used to describe the materials employed in the process of removing oxygen or preventing it from entering the in-package environment of food products subject to undesirable oxidative reactions. Many of the terms appear to be used interchangeably. Each of the commonly used terms is assigned a definition, accepting the fact that various professionals will debate these interpretations. It is also probable that these definitions do not necessarily have clear boundaries, i.e., they overlap.

ANTIOXIDANTS

Antioxidants generally are compounds that react with lipid or peroxide radicals or, in light, with singlet oxygen, and that are themselves oxidized to generate what are generally innocuous nontoxic compounds. Antioxidants are commonly incorporated into the food product itself as contrasted to being included as a part of a package material system. For many years, antioxidants were—and still are—fat-soluble compounds incorporated into fatty foods to preferentially react with intermediate oxidation products in the surrounding air or dissolved or occluded in the food product. In more recent years, the term *antioxidant* has been used more broadly to encompass compounds that react in non-lipid environments, such as, for example, in human body cells, which are water-based environments. The classic lipid antioxidants include such compounds as:

- BHA—butylated hydroxyanisole
- BHT—butylated hydroxytoluene
- PG—propyl gallate

All of these are often blended with lipids to retard their oxidation. The BHA/BHT compounds are also often incorporated into polyolefin package films to retard the oxidation of the plastic materials. Yet another application for the lipid antioxidants is incorporation into flexible package materials for intentional diffusion to the surface, sublimation into the in-package environment, and subsequent incorporation into dry food products such as breakfast cereals, where the actual antioxidant effect takes place. Among the limitations of such systems is that they act only against lipid oxidation, i.e., they exert relatively little action in moist environments, although they can function in emulsions such as salad dressings. Further, they are slow and the reactions are so limited that very large quantities are required to function as effective antioxidants.

More recent additions to the list of package-material antioxidants are beta carotene, a vitamin A precursor, and alpha tocopherol (vitamin E). The vitamin A precursors are claimed to be useful in retarding oxidation in polyolefins. No claims have been made to date on their ability to reduce the passage of oxygen into the packages or to remove oxygen from the interiors of packages.

Among the other products classified as antioxidants for incorporation into package structures are ascorbic acid and its analogues. Such compounds are water soluble and effective in moist environments. In many products, ascorbic acid is called an oxygen interceptor. In the context of packaging, ascorbic acid is usually classified as an oxygen scavenger.

OXYGEN INTERCEPTORS

Interceptor is a somewhat vague word describes any compound that operates early and actually blocks the adverse effect of oxygen in the air on the food, before the oxygen can enter the food. The word *interceptor* has been used to describe compounds in both the product and package materials. Often the word *interceptor* has been used as a descriptor on food labels to avoid statements about antioxidants that may convey images of undesirable chemical reactions to some consumers.

Interceptors prevent oxygen from reaching the food product by themselves being oxidized before the oxygen reacts with the food product. *Interceptors* are sometimes erroneously regarded as synonymous with *antioxidants*.

OXYGEN ABSORBERS

Technically, absorbers remove oxygen by physically trapping the oxygen and not through chemical reaction. In practice, however, there are very few, if any, materials that can effectively remove oxygen by physical as contrasted with chemical reaction. It appears that the word *absorber* is often used loosely to describe any system that removes oxygen from the environment to delay or prevent oxidation reactions in food products.

OXYGEN SCAVENGERS

The term *oxygen scavenger* has been applied to materials incorporated into package structures that chemically combine with, and thus effectively remove, oxygen from the inner package environment. In turn, scavengers may remove oxygen from the food product itself through diffusion resulting from differential partial pressure actions.

One implication that might be derived from this list of definitions is that scavengers are fast-acting, high-capacity oxygen interceptors (if the scavengers are operating at the package gateway), capable of eliminating relatively large volumes of oxygen and continuing their action indefinitely as long as the scavenger is present.

The most widely known or proposed commercial oxygen scavengers are ferrous compounds, catechol, ascorbic acid and its analogues, ligands, oxidative enzymes such as glucose oxidase, unsaturated hydrocarbons and polyamides.

RATIONALE FOR IN-PACKAGE OXYGEN SCAVENGING

As much as oxygen is necessary for life, it is also an element that is a major factor in food-product degradation. Oxygen is involved in anabolic reactions such as respiration and is also very much a part of catabolic or deteriorative reactions of foods.

The reactions of oxygen in catabolysis are integral to the deterioration of foods and beverages. Since most food and beverage products are biological in origin, it follows that oxygen was involved in their beginnings in some way and that they ultimately revert back to their components through oxidative reactions.

Although several technologies for actively removing oxygen or scavenging oxygen from package interiors have been discussed for more than 60 years, actual reported physical testing dates back only to the 1950s and real commercialization only to the late 1970s. When Japan's Mitsubishi Gas Chemical Company incorporated their Ageless® reduced iron-containing, gas-permeable sachets into larger packages of foods in 1977, active-packaging oxygen scavenging became prominent as a commercial technology to help preserve contained foods.

Much of the effort dedicated to the preservation of foods and beverages is targeted toward elimination or exclusion of oxygen, which would otherwise exert adverse effects. Among the many adverse end effects of oxygen on and in foods and beverages are:

(1) Oxidative rancidity of unsaturated fats leading to off-flavors and even, in extreme circumstances, to toxic end-products
(2) Loss of ascorbic acid or vitamin C, especially in fruit- and vegetable-based foods
(3) Darkening and browning of fresh meat pigments
(4) Fostering growth of aerobic spoilage microorganisms
(5) Staling odors in soft bakery goods
(6) Hatching of insect eggs and growth of insects
(7) Acceleration of fresh fruit and vegetable respiration
(8) Enzymatic and nonenzymatic phenolic browning of fresh fruit flesh

(9) Oxidation of aromatic flavor oils of beverages such as coffee and tea
(10) Flavor deterioration of beer
(11) Discoloration of processed fruit and vegetable pigments

In effect, most foods and beverages are in one or another way adversely affected by the presence of oxygen, which reduces nutritional content or degrades color, flavor, or texture or combinations of these.

FOOD DETERIORATION

Among the vectors of food deterioration are microbiological reactions, often favored by the presence of oxygen to provide the environment for the growth of aerobic or "oxygen-loving" microorganisms; enzymatic reactions, in which biologically catalyzed oxidative reactions require the presence of oxygen; and biochemical reactions, in which uncatalyzed oxidation reactions usually require the presence of oxygen. Thus, oxygen plays a major role in many spoilages and quality losses of foods.

Microbiological Deterioration

Microbiological deteriorations are usually spoilages that take place in foods with high water activity at temperatures above the foods' freezing points. As the temperature increases, the rate of microbiological growth, and hence spoilage, increases exponentially, usually up to the temperature at which the microorganisms may be thermally disrupted or destroyed. Most, but not all, spoilage microorganisms are aerobic and so consume oxygen in their respiratory actions. Thus, the reduction or removal of oxygen from the environment of aerobic microorganisms should reduce their rates of growth and thus the rate of spoilage. This reduced rate of microbiological growth under reduced oxygen conditions is among the reasons for the effectiveness of vacuum and modified-atmosphere packaging for nonsterile foods. Alternatively, complete removal of oxygen can result in conditions under which anaerobic spore-forming microorganisms and particularly pathogenic types are capable of growing and producing toxins. Thus, oxygen reduction to retard microbiological spoilage could lead to an undesirable result in food products whose water activity is sufficiently high to permit microbial growth. A major dichotomy today is that packaging (or other) technologies to remove oxygen in order to retard spoilage have the potential to enhance some pathogenic microbiological activities.

Enzymatic Deterioration

Foods whose enzyme systems have not been inactivated by heat, water activity control, or analogous input are vulnerable to deterioration accelerated by biological catalysts or enzymes. In normal respiration, again in foods with sufficiently high water activity, many enzymatic reactions are oxidations and so require the presence of oxygen. Among the types of reactions in which enzymes are involved are phenolic browning/discoloration, vitamin C deterioration, and flavor loss. Removal of oxygen thus can assist in reducing enzymatic oxidative food deterioration.

Biochemical Deterioration

Even when all biological vectors such as microorganisms and enzymes have been inactivated, nonenzymatically driven biochemical deteriorations can occur in foods, with the rates depending on temperature and moisture content. Usually, the higher the temperature and the higher the moisture concentration, the more rapid the reaction, but reactions can occur even at temperatures below freezing and at moisture contents as low as 0.1%. Among the types of deteriorative biochemical oxidative reactions that can occur are lipid oxidation, vitamin B and C oxidations, nonenzymatic phenolic browning, and flavor oxidations. Thus, in products that are otherwise biologically stabilized for ambient temperature distribution, as in thermally processed foods, oxygen can and does play a role in further quality losses, even leading to unacceptable products.

MEASURES TO EXCLUDE OXYGEN

Recognition of the adverse consequences of the presence of residual oxygen within a food package has led to a wide array of technologies and mechanisms to effectively reduce and/or exclude oxygen:

- hermetic packaging in gas-barrier package structures, including closures
- processing and hermetically packaging under vacuum, e.g., cured meats, roasted and ground coffee
- packaging under inert gas such as nitrogen, e.g., instant coffee, unit-portion aseptically packaged apple sauce, nuts
- deaerating and packaging, e.g., orange juice
- sparging with an inert gas such as nitrogen and hermetically packaging, e.g., instant coffee, edible oil, or solid shortening
- blanketing with an inert gas such as nitrogen and hermetically packaging, e.g., cheese, peanut butter
- deaerating and completely processing and packaging in the absence of air, e.g., aseptic packaging of beverages on a Tetra Pak vertical form/fill/seal system or in the past, a Crosscheck deposit/fill/seal system
- applying a vacuum and tightly sealing, e.g., fresh red meat (which fosters the purple myoglobin color over the red oxymyoglobin color), some cured meats, fresh-cut produce

- displacing the air with a reduced-oxygen-gas environment and heat sealing, e.g., fresh vegetables under modified atmosphere
- tightly sealing the product and permitting the contents to respire and consume the oxygen present, e.g., fresh-cut vegetables, ground meat in chub packages
- heating to expel the air, packaging, injecting steam, followed by hermetic sealing and cooling to condense the steam into water and thus create a vacuum, e.g., most low-acid food-canning processes
- heating to expel the air, filling while hot, and hermetically sealing and cooling to condense the headspace steam vapors, thus creating a partial vacuum, e g., most high-acid beverages and foods such as ketchup
- hot filling, followed by liquid nitrogen injection and hermetic sealing, e.g., fruit and tomato juices, juice drinks, and some soups for ambient-temperature distribution in aluminum cans or polyester bottles
- mechanical vacuum, e.g., cured meats and cheeses in flexible pouches
- mechanical vacuum plus adding in-package oxygen scavenging sachets or labels, e.g., dried beef snacks, jerky, etc.
- removing oxygen, sparging with carbon dioxide, filling headspace with carbon dioxide, packaging in PET bottles containing oxygen scavenger in the bottle walls plus in the closure liner
- blanketing or injecting with inert gas such as nitrogen plus placing in-package oxygen scavenging sachets, e.g., some bakery goods, bacon bits, fresh pasta, etc.
- blanketing or injecting with active gas such as carbon dioxide, plus placing oxygen scavenging sachet, e.g., retail cuts of fresh red meat, soft bakery goods
- adding antioxidant chemicals to the package material, e.g., BHA/BHT to potato chip or dry cereal package materials

Closure measures that are undertaken to exclude oxygen after packaging include:

- double-seam mechanical closure with resilient liner compound to hermetically seal gas-impermeable metal cans
- crown closure for glass bottles with resilient plastic liner to seat on the glass finish for hermetic sealing
- long, moist cork plugs (e.g., wine)
- long, plastic plugs (e.g., wine)
- molded plastic screw closures with resilient plastic liners to seat on glass or plastic bottle finishes (e.g., carbonated beverage bottles)
- roll-on aluminum closures with resilient plastic liners to seat on glass bottle finishes (e.g., beer bottles)
- roll-on aluminum closures with resilient plastic liners to seat on plastic bottles (e.g., carbonated beverages)

- screw-on metal closures with resilient plastic liners to seat on glass, bottle, or jar finishes (e.g., juices)
- clamp-on metal closures with plastisol liners that deform under heat to the shape of the glass jar finish (e.g., baby foods)
- fusion heat sealing of identical plastics (e.g., retort pouches)
- peelable heat sealing of different plastics (e.g., aseptically packaged apple sauce in unit-portion cups)
- mechanical clip bunching of plastic bags (e.g., fresh primal cut meat "barrier bags," chub-packaged ground meat, sausage tubes)
- caulking with heavy layers of flowable waxes or plastic materials (e.g., coated or laminated paperboard folding cartons, such as for unit-portion roasted and ground coffee packages)

Although mechanical oxygen removal is effective in reducing the rate of oxidative deterioration, it is a one-time action. As much of the oxygen as possible is removed in this single action, and the residual oxygen, however low, is present and available to react with the food. When residual oxygen is present it will react with the food, usually in an adverse manner.

OXYGEN ENTRY AFTER CLOSURE

Often a more serious problem occurs with packaged food when oxygen from the air external to the packaged food enters the package:

- by permeation through plastic materials, can-end seam compounds, or glass closure-gasket compounds
- by transmission through heat-seal faults, pinholes, or cracks in aluminum foil and other materials

Thus, almost all hermetically sealed packages are not truly hermetic because they can permit some entry of air and thus oxygen. This entering oxygen can continue any oxidative reactions that have been initiated as a result of the presence of residual oxygen after the packaging operation. Further, some package materials such as polyester may contain dissolved oxygen, and expanded polystyrene usually contains trapped oxygen in the cells.

HISTORICAL BACKGROUND OF OXYGEN ABSORBERS AND SCAVENGERS

Recognition of secondary oxygen reactions began nearly a half century ago. The early research centered first on the knowledge that residual oxygen was a reactant whose removal could bring about superior quality retention in packaged foods. Among the materials incorporated into the food itself to reduce the residual oxygen were ascorbic acid and glucose oxidase/catalase. These materials proved so technically effective that they were incorporated into small gas-permeable flexible pouches or bags placed into the food package, not only to remove residual oxygen, but also to later intercept any oxygen entering the package. These pouches were the predecessors of the present ferrous iron powder sachets.

During the 1970s, concern about oxidation reactions resulting from residual and entering oxygen increased because of the greater use of plastic packaging, all of which is oxygen permeable, and the desire for prolonged quality retention in many packaged foods. A result of these concerns was the development of new means to remove oxygen by chemical reaction using in-package packets or sachets of reactive chemicals and the beginnings of a new subcategory of packaging called *active packaging*. Insight into the current and probable future situation may be gained by a review of the chronology of oxygen-scavenger systems.

EARLY WORK

As early as the 1920s, reports were published on the removal of oxygen from enclosed food components using mixtures containing ferrous sulfate plus moisture-absorbing substances, copper powder and ammonium chloride, or zinc and/or alkali earth metal salts. Research into methods of removing oxygen from closed packages began in earnest during the 1940s in the United Kingdom and was based on the oxidation of ferrous sulfate. None of the very early developments appear to have met with commercial success; retrospective research shows that most appeared to have been technically effective in an *in vitro* context.

Early research led to an American Can Company oxygen-scavenger packaging system applied by the U.S. Army for dry foods, which is still often cited in historic literature as a pioneering effort. This system required packaging the product in metal cans under 8% hydrogen gas (with the remainder nitrogen), which would react with residual and entering oxygen at palladium catalyst sites on the interior of the can lid. At that time, as well as now, air could enter through gaskets in can closures.

An application to flexible pouches occurred when the company's research demonstrated that the reaction of hydrogen with oxygen could be achieved by sandwiching a palladium catalyst in a plastic barrier lamination, in one case with alumina as the support and with a desiccant. In another variation, the need to gas flush was avoided by employing calcium hydride to produce hydrogen on reaction with water vapor from the food; the catalyst and the hydride were in a gas-permeable sachet enclosed within the package. However technically effective this system was, the concept of packaging with hydrogen had hazards, and so the system did not become commercially viable (Warmbier and Wolf, 1976).

Oxygen absorbers using dithionite (sodium hyposulfite and hydrosulfite) as their main component were marketed in Japan during the late 1960s but were never widely accepted because they lacked stability during handling and storage and had some undesirable side reactions. A patent was later issued on this process on an improved system involving dithionite, calcium hydroxide, activated carbon, and water (Bloch, 1965).

1960s AND 1970s

The first major commercial oxygen scavengers, as the active-packaging oxygen-reacting systems came to be called, were from Japan's Mitsubishi Gas Chemical Company. During the 1970s, this company introduced reduced iron salts into oxygen-permeable sachets, which were placed in sealed gas-barrier food packages. Any oxygen within or entering the package oxidized the iron to the ferric state in the presence of moisture drawn from the food product. Under the tradename Ageless®, oxygen-scavenger packets were commercialized in Japan and introduced into the United States during the late 1970s. These systems were soon followed into the commercial market by Japan's competitor Toppan Printing Company with ascorbic acid–based oxygen scavenger systems and later by other Japanese competitors.

Since then, more than a dozen Japanese companies, two American companies, and a Taiwanese company have been actively promoting the use of sachets of readily oxidizable materials as oxygen scavengers for use in flexible and other packages. It has been claimed that the package internal oxygen concentration can be lowered to below 0.0001% with oxygen absorbers when appropriate scavenging materials and barrier package structures are used. Gas flush and analogous physical methods to achieve and maintain oxygen levels below 0.5% have not proven economical. These Ageless® sachets and their iron and non-ferrous analogues have been the basis for some package materials containing oxygen scavengers in commercial use and under development.

During the 1960s and 1970s, the technology of oxygen removal included the incorporation of antioxidants into package materials. Such compounds as propyl gallate or butylated hydroxyanisole in carriers were blended into thermoplastics, ostensibly to reduce the rate of oxygen permeation through gas-barrier plastic materials such as films or semirigid cups, tubs, and trays. The concept was that antioxidants would retard oxidative reactions within the plastic arising from oxygen passage. Among the sources for these concepts were W. R. Grace and Company (Columbia, MD) and the late noted plastics innovator Dr. Emery Valyi, whose focus was on multilayer barrier plastic cups. It is possible that Dr. Valyi's efforts resulted in prototype packages, but no literature reference indicates reduction to commercial practice (Valyi, 1977).

1970s AND 1980s

During the 1970s and 1980s, research on incorporating sulfites into package materials came from bag-in-box systems' developer/manufacturer Scholle Corporation (Northlake, IL) and from American National Can Company. The former was directed toward incorporation of the sulfites into multilayer flexible package materials for the explicit purpose of removing oxygen from the interior of the package. The latter were aimed mainly at removing oxygen from the interior of semirigid multilayer barrier plastic retort cans, often called buckets. As a major producer of bucket-type cans containing moisture-sensitive ethylene vinyl alcohol as the principal oxygen barrier, American National Can Company pioneered in incorporating a desiccant layer into their multilayer plastic structure. The next step was to supplement the desiccant with an oxygen scavenger, which, in this instance, came from the sulfite group. This concept was not commercialized

1980s

Since about 1980, Australia's Commonwealth Scientific Industrial Research Organization (CSIRO) has been developing systems based on the reactions of singlet oxygen. The polymer film is utilized as both a solvent for oxygen and a reaction medium for the scavenging reaction. The technique involves polymer films containing a photosensitizing dye and a singlet-oxygen acceptor. Singlet oxygen is transient single-atom oxygen. On illumination of the film with visible light, excited dye molecules convert oxygen diffusing into the film from the external environment to singlet oxygen. The singlet-oxygen molecules diffuse to react with the acceptor and so are consumed. The process continues as long as the film is illuminated and acceptor is available. Although an advantage of this type of system is that it requires no addition of sachets to the food package, a major disadvantage is that the reaction does not occur in the dark and would begin immediately on storage of the plastic in light (Rooney, 1982; Rooney, 1984). CSIRO (now Food Science Australia) continues to be one of the leaders in developing oxygen-scavenging package materials, although to date, despite the considerable investment, few commercial products have resulted (Rooney, 1993a, 1993b, 1997).

Advanced Oxygen Technologies, formerly Aquanautics Corporation, incorporated organometallic ligand compounds analogous to hemes, which bind oxygen, into the plastic liners inside the bottle cap of beverage and particularly beer containers. The oxygen in the headspace of the glass bottle is removed by the compound, thus prolonging the shelf life of the product. This technology has been commercialized by American bottle-closure producer ZapatA Industries, Inc. (Frackville, PA), and adopted by some breweries.

Aquanautics' research began from a U.S. Navy program to create an artificial gill for underwater "breathing" activities. In mid-1994, Advanced Oxygen Technologies ceased activity and their technology was acquired by W. R. Grace and Company. Some of their proposed applications include oxygen-absorbing spots, lines or coatings on the walls and/or lids of cans, and polymer blends for the flexible packaging of solid foods and closure liners (Zenner and Salamer, 1989; Zenner et al., 1992; Zenner, 1992; Zenner et al., 1993).

In another approach, the United Kingdom's Metal Box, then CMB and Carnaud MetalBox and now Crown Cork and Seal, developed Oxbar™. The active ingredients were MXD6 nylon and a cobalt salt to catalyze the reaction of oxygen with the aromatic nylon. Oxbar™ incorporated these active ingredients into the wall of a semirigid polyester bottle. The plastic retards oxygen ingress into the container and scavenges oxygen from inside the container as well. This system was abandoned by Carnaud MetalBox in 1992 for both economic and regulatory reasons, but the technology appears to have been revived by Continental PET Technologies for its multilayer polyester beer bottle, which was commercialized in 1998–1999 and has been revived by Crown Cork and Seal (Anonymous, 1990; Brody, 1999; Cochran et al., 1989; Cochran et al., 1991; Cochran et al., 1993).

The use of enzymes to scavenge oxygen has been known for many years, having been suggested in the early 1900s. In 1956, patents were issued for sachets containing glucose oxidase/catalase enzymes and also a glucose oxidase–treated cellophane film. Trials were conducted with a glucose oxidase sachet system referred to as OxyBan, but it appears not to have been used commercially (Scott, 1958; Scott, 1965; Scott and Hammer, 1961). In 1989, Pharmacal, an entrepreneurial development company, reported on an interactive package incorporating these two enzymes to remove oxygen, but this system has not been commercialized.

1990s

In 1990 in Japan, can maker Toyo Seikan Kaisha, Ltd., announced the incorporation of an Oxyguard® ferrous oxygen scavenger in their package material coextrusion, in which the scavenger is in the core of a film with an oxygen-permeable film on the interior and an oxygen barrier on the exterior. The mechanism for initiation was not described (Abe, 1990).

Also in 1990, Japan's Fujimoro announced the insertion of gas-permeable oxygen scavenger sachets into the space between the inner and adjacent outer layer of packaging films. Slots in the inner film permit air within the package to pass into the oxygen scavenger sachet.

In 1990, W. R. Grace and Company announced the European commercialization of their Daraform® closure liner incorporating an unidentified oxygen scavenger, which was assumed to be ascorbic acid. In 1991, Grace began marketing this liner in the United States for beverage packaging applications.

Almost simultaneously, what was then the Cryovac Division of W. R. Grace and Company began importing and marketing Mitsubishi Gas Chemical Company's Ageless® reduced-iron–based sachets in the United States. Cryovac's efforts paralleled those of their principal, Mitsubishi Gas Chemical.

At about the same time, another group within Grace independently initiated development of yet another oxygen scavenger technology for packaging—the incorporation of unsaturated hydrocarbons into plastic films. This development, which was commercially realized in 1998 in a product tradenamed "Stealth" or Cryovac® OS1000, is detailed later in this text (Cook, 1999).

In 1991, Multiform Desiccants, Inc., later renamed Multisorb Technologies, Inc., which had been supplying reduced iron oxygen scavenger sachets, announced their development of a flat manifestation of an oxygen scavenger sachet that can be directly affixed to the interior of a package as if it were a label, i.e., by adhesive. The surface area of this label, designated FreshMax®, is large (Idol, 1991, 1993a, 1993b). Also in 1991, Mitsubishi Gas Chemical Company commercialized an oxygen scavenger label that is similar to that produced by Multiform Desiccants. This product was commercialized during the late 1990s. One problem with ferrous-based oxygen scavengers is that moisture is needed to activate the reaction. Moisture-barrier heat-sealing polyolefins on the package interior retard movement of water vapor from the contents to the scavenger.

Japan's Keplon marketed their reduced iron–based oxygen scavenger sachet products in the United States for three years through two different agents. Japan's Toppan Printing has sold their ascorbic acid–based oxygen scavenger sachets in the United States. An American company, United Desiccants Gates (now Sud-Chemie), marketed Japan's Nippon Soda's iron-based oxygen scavenger products under the OxySorb name. These thrusts reflect a distinctly Japanese base to much of the oxygen scavenger commercialization to date.

COMMERCIAL APPLICATIONS

The following are commercial applications of oxygen scavengers in the United States:

- modified-atmosphere–packaged fresh-chilled pasta—reduced iron contained in sachets adhesively attached to the interior base trays
- dry beef snacks (retail packs and export packs to Japan)—reduced iron in sachets, for beef jerky or pemmican
- precooked poultry master packs (distribution pack)—reduced iron in sachets

- real bacon bits (glass jar)—reduced iron in sachets
- macadamia nuts in jars—reduced iron in sachets
- pepperoni chips (both retail and distribution pack)—reduced iron in sachets
- fresh red meat distribution package (activated by injection of acetic acid solution)—reduced iron in sachets (DelDuca, 2000)
- bulk peanuts (distribution pack)—reduced iron in sachets
- bakery goods
  - specialty "health" non-wheat (i.e., gluten-free) bread—reduced iron in sachets
  - soft pretzels—reduced iron in sachets
- regenerated collagen sausage casings (distribution package)—reduced iron in sachets
- polyester beer, juice and ketchup bottles—incorporated into the core layers of polyester
- closure liners for beer bottles—incorporated into the plastic liner
- a few pharmaceuticals and medical supplies
- candy
- several shelf-stable U.S. military ration items (i.e., bread, cake)—reduced iron in sachets (Pruskin, 1996)
- NASA (shelf-stable tortillas)—reduced iron in sachets

Although quantities of dried beef snacks containing in-package oxygen scavenger sachets are distributed in the United States, many such packages are for export to Japan.

One manufacturer of precooked poultry employs in-package oxygen scavenger sachets for its master or distribution packs containing prepackaged product.

U.S. meat packer Hormel employs in-package oxygen scavenger sachets in the bottom of retail glass jars of ambient-temperature, shelf-stable real bacon bits as well as in a flexible pouch of pepperoni sausage slices.

Kraft General Foods' DiGiorno® brand fresh-chilled pasta uses "zero oxygen" with in-package oxygen scavenger sachets adhered to the bottom interior of the tray.

Collagen sausage casings are made from animal collagen and extruded as uniform flexible tubes to be used as natural casings, mostly for small pork breakfast sausage. They are packaged using oxygen scavenger sachets. This product represents only a small fraction of the total U.S. sausage-casing market.

The current market for in-package oxygen scavenger sachets in the United States, although it had been narrow and specialized, had grown from about 100 million units in 1995 to about one billion units by the late 1990s.

In 2000, about 500 million polyester bottles of beer juice and ketchup incorporated oxygen scavengers. In-package oxygen scavenger sachets are well known in the United States and have been tested for many applications. In fact, in several situations, in-package oxygen scavenger sachets were used commercially and later withdrawn, e.g., in cans of roasted and ground coffee because they evidently did not have sufficient oxygen to remove.

UNITED STATES MILITARY APPLICATIONS

Outside Japan, and other than for use with fresh pasta and in the dried meat sector, ferrous-based oxygen scavenger sachets are primarily used for U.S. military shelf-stable ration components. During the late 1980s, oxygen scavenger sachets were introduced first into packages of water activity–controlled, shelf-stable pouch bread to provide protection against mold growth. In the absence of the oxygen scavenger sachet, growth of *Aspergillus* and *Penicillium* was visible on the bread within 14 days (Powers and Berkowitz, 1990). With oxygen scavengers, up to three years' shelf life at 27°C (80°F) can be achieved while preserving flavor, texture, and nutritional value.

With shelf stability being augmented through modified-atmosphere packaging, introduction of the pouch bread has resulted in a variety of shelf-stable baked goods for the military that are incorporated into field rations. White and whole wheat bread is used to supplement the Meal, Ready-to-Eat (MRE) and the Unitized Group Ration (UGR). The MRE is the standard combat ration used by the Armed Services to sustain individuals during operations that preclude organized food-service facilities, but in which resupply is established or planned. The UGR provides high-quality group meals in the field by integrating components of heat-and-serve rations with quick-prepared and/or ready-to-use brand-name commercial products. It is used to sustain groups of military personnel during operations that allow organized food-service facilities.

Iron-based oxygen scavenger sachets are now contained inside numerous dry and intermediate-moisture packaged rations, such as:

- white and whole wheat bread, MRE
- pound cakes, MRE (7 varieties)
- snack items, MRE
- wheat snack bread, MRE
- fudge brownies, MRE
- hamburger buns, UGR
- waffles, UGR

All of these military rations are shelf-stable, which is why they are ready to eat as is.

Table 4 indicates the approximate 1999 procurement levels for ration components packed with oxygen scavenger sachets.

TABLE 4. 1999 U.S. Military Procurement Quantities (All Services).

Item	Total Individual Packages
White bread	2,880,000
Whole wheat bread	480,000
Pound cakes (6 varieties)	7,523,474
Fudge brownies	2,405,050
Wheat snack bread	2,340,000
Potato sticks	1,250,000
Chow mein noodles	1,250,000
Nut raisin mix	1,250,000
Pretzels	1,250,000
Waffles	14,800
Hamburger buns	71,000
Total	*20,714,324*

The bakery products for the U.S. military are currently produced and packaged by Sterling Foods, Inc., San Antonio, Texas. Current vendors and/or assemblers for MRE snack items include: The Wornick Company, McAllen, Texas; Southern Packaging & Storage Company (SOPAKCO), Mullins, South Carolina; Ameriqual Foods, Inc., Evansville, Indiana; and Trans-Packers Services Corporation, Brooklyn, New York (co-packer). The UGR waffles (plain and blueberry) are produced by DeWaffelbakkers, S. Little Rock, Arkansas (Kline, 1999).

The same objections associated with the use of oxygen scavenger sachets in commercial applications can be applied to their use in military rations. These concerns include the risk of the product manufacturer overlooking insertion in the package, the potential for the sachet being broken, and the risk of the consumer accidentally ingesting the sachet.

JAPANESE SITUATION

More than two billion oxygen-scavenger sachets are used annually in a growing Japanese market, led by Mitsubishi Gas Chemical Company, which holds about 70% of the Japanese market. At least eleven Japanese companies have competed in the oxygen scavenger market, although not all are currently offering oxygen scavengers (Table 5). Most, but not all, of the oxygen scavengers in Japan are based on the reaction of ferrous iron with oxygen in the presence of moisture.

In-package ferrous-based sachets may be classified as self-working, i.e., those that contain moisture in sachets and start oxygen absorption when

TABLE 5. Japanese Oxygen-Scavenger Manufacturers.

Manufacturer	Tradename	Year Entered the Market
Mitsubishi Gas Chemical	Ageless®	1977
Keplon	Keplon™	1978
Oji Kako	Tamotsu™	1979
Toppan	Freshilizer™	1980
Toagosei Chemical Industry	Vitalon™	1980
Nippon Soda	Secule™ (OxySorb™ in the United States)	1980
Hakuyo	Sansoless™	1980
Finetec	Sansocut™	1986
Nippon Kayaku	Modulan™	1987
Powdertec	Wonderkeep™	1987
Ueno Seiyaku	Oxyeater™	1988
Dai Nippon	Sequl®	1993

exposed to air, or as moisture-dependent, those that contain no moisture in the sachets but extract the moisture from the in-package atmosphere. Non-iron–based scavengers include catechol-based scavengers. Table 6 indicates the types of in-package oxygen scavengers in commercial use in Japan in approximate decreasing order of commercial importance, the conditions of use, and typical applications.

The major applications reported in Japan are in packaging the following:

- bakery goods, to prevent mold, e.g., soft cakes, muffins, bread, English muffins, biscuits (including chocolate coated), pizza, and cheesecake
- dry snacks, to reduce lipid rancidity
- nuts
- chocolate-coated nuts
- beef jerky
- processed seafoods
- seasonings
- Japanese dry noodles
- cheese
- aseptically packaged cooked rice
- seaweed flakes
- dry pet foods
- ophthalmic tablets
- cough capsules
- plant growth hormone
- antibiotics

TABLE 6. Classification of Oxygen Absorbers in Japan.

Function	Reacant	Typical Applications	Absorption Speed	Japanese Product	Manufacturer
$O_2\downarrow$	Iron	Self-working	Dry; $a_w < 0.3$; tea, nuts	4–7 days	Ageless® Z-PK Vitalon™ T	Mitsubishi Toagosei
			Medium a_w ($a_w < 0.65$); dried beef	1–3 days	Ageless® Z Keplon™ TS	Mitsubishi Keplon
			High a_w ($a_w > 0.65$); cakes, bakeries	0.5 day	Ageless® S Secule™ CA	Mitsubishi Nippon Soda
			Frozen; +3–25°C; raw fish	3 days at −25°C	Ageless® SS	Mitsubishi
		Moisture-dependent	High a_w ($a_w > 0.85$)	0.5 day	Ageless® FX	Mitsubishi
	Catechol	Pastas Self-working	Medium a_w ($a_w < 0.65$); nuts		Vitalon® LTM Tamotsu™ A	Toagosei Oji Kako
			High a_w; ($a_w > 0.65$); cakes		Tamotsu™ P	Oji Kako
$O_2\downarrow$ & $CO_2\downarrow$	Iron and calcium	Self-working	Roasted/ground coffee	3–8 days	Ageless® E	Mitsubishi
$O_2\downarrow$ & $CO_2\uparrow$	Ascorbic acid	Self-working	Medium a_w ($a_w > 0.3$), (<0.5); nuts	1–4 days	Ageless® G	Mitsubishi
	Organic acid and iron	Moisture-dependent	High a_w ($a_w > 0.85$); cakes		Toppan™ C Vitalon™ GMA	Toppan Toagosei
$O_2\downarrow$ & ethanol $\uparrow$	Iron and ethanol on zeolite	Moisture-dependent	High a_w ($a_w > 0.85$); cakes		Negamold™ (antimycotic)	Toppan

- chestnuts
- anti-lice carpet powder
- natural beta carotene
- vitamin pills/tablets
- medical kits to preserve reagents
- kidney dialysis kits
- pet foods
- feeds
- coffee and tea

In-package oxygen scavenging is used in a broad diversity of applications, none of which is large by itself. There do not appear to be mainstream applications in which all products within the category employ a scavenger product. In Japan, in-package oxygen scavenger sachets are specialty components used with various foods, pharmaceuticals, and medical kits to preserve reagents. Many product applications are in circumstances in which the product is distributed in extra packaging since it is intended as a gift. Product turnover is therefore slow, and the time in distribution is relatively long. Thus, there is a need for a means to prolong quality retention.

BIBLIOGRAPHY

Abe, Y. 1990. "Active Packaging—A Japanese Perspective." *Proceedings of the International Conference on Modified Atmosphere Packaging*, Stratford-upon-Avon, United Kingdom. Sponsored by Campden Food and Drink Research Association, Chipping Campden, United Kingdom. October 15–17.

Anonymous. 1990. "Oxbar: The Oxygen Barrier." *Food Engineering International*, 15(8):51.

Bloch, Felix 1965. "Preservative of oxygen-labile substances, e g , foods." U S. Patent 3,169,068. February 9.

Brody, A. L. 1999. "Oxygen Scavenging Packaging: Where Are We? Where Do We Go From Here?" *Proceedings of "Oxygen Absorbers: 2000 and Beyond" Conference*, Chicago, Illinois, George O. Schroeder Associates, Inc., Appleton, Wisconsin. June

Cochran, Alexander, Rickworth Folland, James William Nicholas, and Melvin E. R. Robinson. 1989. "Improvements in and relating to packaging." European Patent Application 88306175.6, Publication number 0 301 719. February 1.

Cochran, Michael A , Rickworth Folland, James W. Nicholas, and Melvin E. R. Robinson. 1991. "Packaging." U.S. Patent 5,021,515. June 4.

Cochran, Michael A., Rickworth Folland, James W. Nicholas, and Melvin E. R. Robinson. 1993. "Process for production of a wall for a package " U.S. Patent 5,239,016. August 24.

Cook, P. 1999 "Stealth Scavenging Systems: A Remarkable Breakthrough Technology." *Proceedings of "Oxygen Absorbers: 2000 and Beyond" Conference*, Chicago, Illinois, George O. Schroeder Associates, Inc , Appleton, Wisconsin. June.

DelDuca, Gary. 2000. "Innovative Packaging Concepts for Meat and Poultry" Institute of Food Technologists' Annual Meeting Presentation June.

Idol, R. C. 1991. "A Critical Review of In-Package Oxygen Scavengers." *Sixth International Conference on Controlled/Modified Atmosphere/Vacuum Packaging*, Schotland Business Research, Inc , Princeton, New Jersey.

Idol, R. C. 1993a. "A Retail Application for Oxygen Absorbers in Europe." *Pack Alimentaire '93*, Schotland Business Research, Inc., Princeton, New Jersey.

Idol, R. C. 1993b. "Oxygen Absorbing Labels." *Packaging Week*, 9:16.

Kline, L. 1999. "Oxygen Absorber Technology for U.S. Military Applications." *Proceedings of "Oxygen Absorbers: 2000 and Beyond" Conference*, Chicago, Illinois, George O. Schroeder Associates, Inc., Appleton, Wisconsin. June.

Powers, E. M., and D. Berkowitz. 1990. "Efficacy of an Oxygen Scavenger to Modify the Atmosphere and Prevent Mold Growth on Meal Ready-To-Eat Pouched Bread" *Journal of Food Protection*, 53(9):767–770.

Pruskin, Lauri R. 1996. "Oxygen Absorbers Take Active Role in Quality Packaging, Parts I and II." *Packaging Technology & Engineering*, 5(7) and 5(8). July and August.

Rooney, M. L. 1982. "Oxygen Scavenging from Air in Package Headspaces by Singlet Oxygen Reactions in Polymer Media." *Journal of Food Science*, 47(1):291–298.

Rooney, M. L. 1984. "Photosensitive Oxygen Scavenger Films: An Alternative to Vacuum Packaging." *CSIRO Food Research*, 43:9–11. March.

Rooney, M. L. 1993a. "Oxygen scavenging compositions" PCT Patent Application PCT/AU93/00598.

Rooney, M. L. 1993b. "Oxygen scavengers independent of transition metal catalysts." International Patent Application PCT/AU 93/00598. November 24.

Rooney, Michael L. (ed.) 1995. *Active Food Packaging*. Glasgow, UK: Blackie Academic & Professional.

Rooney, Michael L. 1997. "Active Packaging Materials for Modified Atmosphere Generation and Maintenance." Prepared for Institute of Packaging Professionals' *MAPack '97 Conference*.

Scott, Don. 1958. "Enzymatic Oxygen Removal from Packaged Foods." *Food Technology*, 12(7):7.

Scott, Don. 1965. "Oxidoreductase." *Enzymes in Food Processing*. New York: Academic Press.

Scott, Don and Frank Hammer. 1961 "Oxygen Scavenging Packet for In-Packet Deoxygenation." *Food Technology*, 15:12

Valyi, Emery. 1977. "Composite materials." U.S. Patent 4,048,361. September 13.

Warmbier, H. and M. Wolf. 1976. "Miraflex 7 Scavenger Web" *Modern Packaging*, 49:38, 40–41.

Zenner, B. D. 1992. "Smart Cap and Beyond: The Application of Active Oxygen Absorbing Packaging Materials to Foods and Beverages." *Europack '92*.

Zenner, B. and M. Salame. 1989. "A New Oxygen-Absorbing System to Extend the Shelf-Life of Oxygen Sensitive Beverages." *Europack '89*, The Packaging Group, Miltown, New Jersey.

Zenner, Bruce, E. De Castro, and J. P. Ciccone. 1992. "Methods, compositions and systems for ligand extraction." U.S. Patent 5,096,724. March 17.

Zenner, Bruce D , Fred Teumac, Larrie Deardurff, and Bert Ross. 1993. "Amino carboxylic acid compounds as oxygen scavengers" U.S. Patent 5,202,052. April 13.

CHAPTER 3

Oxygen Scavenger Systems

ANTIOXIDANTS

TRADITIONAL antioxidants applied to delay lipid oxidation by reacting with intermediate products of lipid oxidation, i.e., by interrupting free radical chain reactions, have been incorporated into package materials for many years. Rho et al., (1986) showed that coating the interior surface of a polyethylene film package with t-butylhydroquinone (TBHQ) extended the shelf life of contained deep-fried noodles. Technion-Israel Institute of Technology's Dr. Joseph Miltz and his co-workers (1988) demonstrated that when butylated hydroxy toluene (BHT) was incorporated into high density polyethylene film, the shelf life of a contained oat cereal was increased. These researchers asserted that the applications of the antioxidants into the plastic film of the package were more effective than adding the same antioxidant directly into the contained food products.

A 1969 patent assigned to W. R. Grace and Company, an early participant in oxygen scavenging package material structures, described the incorporation of lipid antioxidant n-propyl gallate into both flexible and thermoformable plastic package materials. The development was for polyolefins such as polyethylene and polypropylene and especially for polyvinylidene chloride (PVdC) (saran) film. The antioxidant could be incorporated into twoply or three-ply structures. An example would be two layers of PVdC sandwiching the antioxidant. Also cited as antioxidants were butylated hydroxy anisole (BHA), BHT, and dihydroguaretic acid, among others, with the preference being for n-propyl gallate.

The antioxidants were incorporated on the interior surface of the film from a liquid solution in which the solvent was propylene glycol, glycerol, or edible oil. The antioxidant in the solvent was trapped between the two film layers.

The antioxidant/solvent mixture between the layers was at a concentration of about 1 to 30 mg/ft.[2] This structure was intended to reduce the oxygen permeability of the film by a factor of up to six. Product contents for the films included cheese. No mention was made of scavenging oxygen from the package interior, but it may be assumed that such a phenomenon occurred if peroxides were formed in the film.

This patent is one of the very few citing conventional antioxidants and is among the earliest publications describing the incorporation of antioxidants into a multilayer film structure, as opposed to into a sachet. Because of the liquid coat, it is not unlike the later Scholle patent (1977), but the Grace patent indicates the layer is very thin and not really in liquid form.

Valyi (1977) indicated the employment of a "getter" as the antioxidant in a semirigid multilayer barrier plastic cup structure. The term "getter" was occasionally used during the 1970s and 1980s to describe a material capable of binding, absorbing, or adsorbing unwanted liquid or gaseous materials. Valyi suggested a multilayer structure of improved oxygen barrier performance as a consequence of the multiple materials functioning together. The layers proposed included an exterior high-barrier plastic and an interior lower-gas-barrier plastic with the getter uniformly dispersed within or between the two layers. Bonding was by adhesive or electrostatic means. The objective was to reduce the oxygen permeability of the total structure. The antioxidant getters cited included BHA and propyl gallate. Valyi's innovation, highly publicized at the time, focused primarily on thermoformable multilayer barrier plastics with separable components, i.e., avoiding the use of coextrusion to achieve the structure. Thus, the "getter" was really a very minor constituent and so this was a secondary issue.

SULFITES

Sulfites and their analogues have been employed as oxygen scavengers for many decades. A very early Carnation Company patent (U.S. 2,825,651) (Loo and Jackson, 1958) described the use of sulfite salt with copper sulfate as a catalyst for oxygen scavenging. The sodium sulfite and its adjunct palletized powders were placed in a separate sachet. The mix removed 8.5 cc of oxygen in 5 hours. If spread into a film, the particles could react without the presence of water because of their intimate contact with the food. In addition, since they could be ground to a fairly fine particle size, they should be able to be incorporated into a film.

Yoshikawa et al. (1977) showed that a similar sulfite-based scavenger system could remove 13 cc of oxygen in 10 hours. The promoters could be metals other than copper, including tin, lead, chromium, nickel, cobalt, palladium, and platinum. Their patent claimed their product could be used for prevention of oxidation of fats and oils, dried foods, raw and fresh foods of all categories, and pharmaceuticals. Their claims were for the composition of the scavenger and not for the method of employment, thus indicating no extension of their product directly into package materials.

In 1965, proposals were made to introduce bisulfites into in-package sachets to consume free oxygen and thus scavenge oxygen from packages of dehydrated eggs, dehydrated whole milk, dehydrated potatoes, edible fats and oils, nuts, and other foods (Bloch, 1965). The compound proposed was sodium bisulfite. One molecule of the bisulfite could consume an atom of oxygen, forming a bisulfate. Metabisulfites were also proposed as alternatives to the sulfites

By itself, a bisulfite does not absorb oxygen to a great extent, and so the surface area of the bisulfite was effectively increased by incorporation into a carrier with a large surface area such as activated carbon or silica gel. To further increase the effectiveness of the bisulfite, a heavy metal activator was proposed with iron salts such as ferrous and ferric sulfates and chlorides. The scavenger could not be in physical contact with the product, and so there had to be a receptacle through which the gases could diffuse. Oxygen removal is not instantaneous but is rapid at the outset, decreasing as the amount of residual oxygen is decreased.

In 1977, the late William Scholle, at the time head of the Scholle Corporation, received a patent on a multiwall flexible pouch material containing a sulfite compound to react with oxygen. In this situation, the sulfite is incorporated into the package in the form of a liquid trapped between sheets of flexible package materials. The fluid itself may be sulfite, bisulfite, metabisulfite, or hydrosulfite. Further, although the exterior ply has low oxygen permeability, the liquid core is intended to react with any oxygen entering from the outside environment, thus preventing it from entering the package interior (Scholle, 1977). The package was intended for use for packaging liquids such as wine, tomato paste, and tomato ketchup.

Any oxygen scavenger producing an end-product compound such as sulfur dioxide is viewed with concern since residuals of this compound can exert an allergic effect on a small proportion of the population.

In 1980, the United Kingdom's Metal Box Company was granted a patent for the use of sodium metabisulfite plus sodium carbonate in a wine bottle bung or cork, in order to release sulfur dioxide to function as an oxygen scavenger (Throp, 1980). The cork or bung was formed by molding ethylene vinyl acetate (EVA) copolymer using an injection molding process in which sulfur dioxide, carbon dioxide, and water vapor are produced to fill voids within the EVA material. This residual SO_2 and water vapor trapped in the voids react with entering oxygen:

$$2 SO_2 + O_2 + 2 H_2O \rightarrow 2 H_2SO_4$$

The most interesting aspect of this patent is the notion of incorporating the oxygen scavenger into the plastic resin before forming it and having the oxygen scavenger actuate as a result of the energy of the forming process.

Dr. Boh Tsai and his associates at American Can Company received patents on 'the incorporation of sulfite oxygen scavengers into plastic structures (Farrell and Tsai, 1987). Dr. Tsai's American Can Company patents were aimed at the company's multilayer retortable barrier plastic cans, often referred to as buckets. In cans such as those fabricated by American Can Company and its successor company, American National Can Company (now known as Pechiney Plastics), the multiple layers of polypropylene, ethylene vinyl alcohol copolymer as the oxygen barrier, and tie or adhesive layers, are coinjection blow-molded into the requisite bowl or bucket shape. The converter also includes a desiccant layer, i.e., a layer of plastic containing a desiccant to preferentially absorb water vapor that would otherwise interfere with the oxygen scavenging capabilities of the moisture-sensitive ethylene vinyl alcohol. In prior developments of oxygen scavenging packages, the scavenger, without complete protection, would react directly with oxygen in the air and thus be ineffective. In this concept, the moisture that passes from the product into the package material during processing is the triggering agent.

The scavenger listed in the American Can Company patent is potassium sulfite, a deliquescent salt that reacts with oxygen only in its wet state. The potassium sulfite may be used by itself or in conjunction with other deliquescent salts that may be actuated at lower relative humidities. Potassium sulfite is cited as an oxygen scavenger that can be readily triggered by the moist high temperature of the retorting process but has enough thermal stability to pass unchanged through thermoplastic processes.

Typical flexible packages consist of inner and outer layers of polyolefins, such as polypropylene, tied to a core ethylene vinyl alcohol layer. The potassium sulfite could be a dispersion in a high density polyethylene carrier placed next to the inner polyolefin material. Although aimed at retortable cans, trays, and tubs fabricated from a total semirigid plastic structure, nothing in the concept would preclude a flexible package material.

Food Science Australia, formerly the Commonwealth Scientific and Industrial Research Organization (CSIRO), has been in the forefront of oxygen-scavenging package materials for several years. Information from this organization has been discussed above; more is provided later in this chapter.

A 1993 patent assigned to W. R. Grace and Company discussed the use of isoascorbates plus sulfites as oxygen scavengers in the plastic gasket liners of bottle closures. The concept in this patent was to remove oxygen from the headspace of glass or even plastic bottles and to significantly reduce or even eliminate the quantity of oxygen otherwise permeating through the liner compound. Sulfites are cited as one of several oxygen scavenger systems that could be incorporated into the closure liner, with no elaboration. Tannins, also indicated by Mitsubishi Gas Chemical Company as catalysts in the 1,2-glycol development, are described in this patent (Speer et al., 1993). Perhaps the relevance of this patent is the notion of direct incorporation of the dry oxygen scavenger in a thermoplastic material.

BORON

A 1992 Mitsubishi Gas Chemical Company patent (Sugihara et al., 1992) described an oxygen absorber that overcomes the problem of triggering metal detectors on packaging lines, which occurs with iron compounds. The oxygen absorber is a reducing boron and an alkaline compound plus a carrier, all contained in an in-package sachet. Suggested are boron itself, boric acid, and salts of boric acid plus alkaline compounds such as sodium carbonate. The proposed carriers are activated carbon and diatomaceous earth. The dry materials are blended and packed in a sachet for placement in a package. The inventors indicate effectiveness in packages of food such as rice. Oxygen concentration within a package could be reduced from the 20.9% in air to less than 0.1% in three days. Tests indicated that the sachet contents could not be detected in a metal detector.

GLYCOLS AND SUGAR ALCOHOLS

Another patent from Mitsubishi Gas Chemical Company (Sugihara et al., 1993) proposed a range of nonferrous compounds in the role of oxygen scavengers.

- 1,2-glycols such as ethylene or propylene glycol, alkaline substance, and a transition metal compound such as iron, cobalt, or nickel halides and sulfates
- 1,2-glycols, alkaline substance, and a phenolic such as catechol or tannic acid or quinone such as a benzoquinone compound functioning as a catalyst
- glycerine and an alkaline substance
- sugar alcohol such as sorbitol, xylitol, or mannitol

These materials, also used to avoid the use of metals that might trigger metal detectors, are blended and placed in gas-permeable sachets to be inserted into high-gas-barrier packages. Among the product contents packaged using this oxygen scavenger system are brown rice and jam-filled sweet soft bakery goods. Neither speed nor capacity of absorption was great compared with alternative scavengers. The significance of this patent was in the identification of 1,2-glycols and particularly propylene glycol, and sugar alcohols such as sorbitol and mannitol as oxygen absorbers.

UNSATURATED FATTY ACIDS AND HYDROCARBONS

In 1978, American pharmaceutical company E. R. Squibb was granted a patent on the use of drying oils such as linseed oil (highly unsaturated) as an oxygen absorber in a gas-permeable "package." The highly unsaturated glyceride was soaked into a porous plastic sponge, which was in turn placed within a protective gas-permeable capsule into which air could flow. A quantity of linseed oil ranging from 0 002 to 0.02 cc per cubic centimeter of air was indicated as sufficient to react with all the oxygen within the container, generally bottles or preformed plastic blisters with heat-sealed aluminum foil closures. Japan's Mitsubishi Gas Chemical Company and others have adopted the concept into several of its sachet scavenger patents.

Focusing on oxygen absorption from within packages of products such as metals or electronic parts subject to corrosion or photographs if exposed to moisture, a 1994 Mitsubishi Gas Chemical patent (Inoue et al., 1994) described the use of unsaturated fatty acids or unsaturated linear hydrocarbons as the oxygen reactants. Such compounds are described as having "remarkably high oxygen absorption capability" over prolonged periods. As in so many other of Mitsubishi Gas Chemical's developments, the oxygen scavengers are placed in gas-permeable sachets, which, in turn, are within the low-gas-permeability food package. The preferred components of the oxygen absorbents were hydrocarbons such as isoprene, butadiene, and squalene. Catalysts such as iron and cobalt salts may be incorporated, and carriers such as activated carbon and zeolite were employed. Data presented indicated that the times needed to attain a 0.1% oxygen concentration in the package were in the 9- to 15-hour range. The absorbers are capable of continuing their activity for at least through 74 days, evidently at ambient temperature. No indication was given on the oxidation products of unsaturated fatty acids, which are, of course, generally malodorous.

The origins of the use of unsaturated fatty acids and their analogues for Mitsubishi Gas Chemical Company were expressed in an earlier patent covering the oxygen absorbent itself (Inoue and Komatsu, 1990). The compounds cited were unsaturated fatty acids and their analogues such as oleic, linoleic and linolenic, and arachidonic acids, contained in oils such as linseed and tung. Transition salts of metals such as iron, cobalt, and nickel were suggested to accelerate the oxidation, with iron preferred. The unsaturated fatty acid was blended with basic compounds such as carbonates or bicarbonates to capture the unpleasantly odorous compounds formed by the reaction between the fatty acid and oxygen. Concentration of oxygen within a pouch could be reduced to less than 0.1% within 24 hours.

Another of the several developments suggesting the application of unsaturated linear hydrocarbons as oxygen absorbers in packaging was in a series of patents assigned to W. R. Grace (Speer and Roberts, 1994). The invention was intended for use in bottle-closure liners and also in coextruded films, preferably for refrigerated foods at low temperatures, especially below 10°C.

The oxygen absorber is an ethylenically substituted or unsubstituted hydrocarbon. An unsubstituted ethylenically unsaturated hydrocarbon has at least one aliphatic carbon:carbon double bond and is exclusively carbon plus hydrogen. A substituted compound has a double bond but is less than 100% carbon plus hydrogen. The preferred hydrocarbons have three or more unsaturated groups. The unsubstituted hydrocarbons cited include isoprene, butadiene, styrene butadiene, oligomers such as squalene, and even beta carotene. The preferred substituted hydrocarbons include esters, carboxylic acids, aldehydes, ketones, and unsaturated fatty acids.

For making transparent oxygen scavenging film layers, 1,2-polybutadiene is preferred, particularly because it can process like polyethylene and because it retains its transparency and physical properties even after almost all of its oxygen capacity has been consumed. Further, it has a relatively high oxygen capacity and a relatively high oxygen scavenging rate, especially at low temperatures because of its low glass transition temperature.

As is apparently required of most higher molecular weight oxygen scavengers, a transition metal catalyst such as iron, cobalt, or nickel, with cobalt preferred, is required in the form of 2-ethylhexanoate or neodecanate.

The scavenging compounds are blended with what are referred to as diluents, which are thermoplastics including polyester, polyethylene, polypropylene, polystyrene, and ethylene vinyl acetate, i.e., all the common packaging thermoplastics. Films may be fabricated by most of the common plastics-fabrication procedures such as coextrusion, solvent casting, and/or extrusion coating, and more rigid containers by injection molding and/or stretch blow-molding.

Scavenger compounds are incorporated in the range of less than 5% of the total weight of the package material. The finished plastic may be a semirigid sheet for thermoformed trays, tubs, and cups or as in extrusion or injection blow-molded bottles.

A typical multilayer film described is an outer low-oxygen-permeability layer (1 to 10 cc O_2/m^2/day at 25°C)/oxygen scavenging layer/inner high-gas-permeability layer. Multilayer flexible film structures may be prepared by coextrusion, coating, or lamination.

The published patents suggested an oxygen scavenging rate of 0.5 cc per gram of scavenger per day at ambient temperature, with capability of even greater than 5 cc per gram per day. Among the examples cited was a multilayer blown coextruded 0.003-inch film consisting of 0.001- to 0.0015-inch ethylene vinyl acetate and a second layer of 90% syndiotactic 1,2-polybutadiene plus the cobalt neodecanoate and a benzophenone photoinitiator. The film is capable of scavenging oxygen at a rate of 450 cc per square meter per day at room temperature. This and succeeding patents are possibly the basis for the Cryovac® Stealth OS1000 oxygen scavenger film introduced in 1998 (Cook, 1999).

Chevron Chemical's oxygen scavenging technology is based on a reaction designed to remove oxygen without producing odorous end compounds. Although many polymers oxidize and therefore could be used as scavengers, they degrade, receiving an oxygen molecule and splintering into smaller degraded components. These smaller molecules, often with adverse sensory effects, are mobile in the polymer matrix and will frequently migrate into the packaged contents, causing adverse sensory results. Controlling initiation of scavenging has been a problem with some scavenging systems.

Chevron's new polymer reportedly readily scavenges oxygen without degrading into smaller, undesirable compounds. Further, with a photoinitiator and catalyst system, the polymer can remain in a nonscavenging state until triggered by UV radiation.

The Chevron terpolymer system has been demonstrated to absorb oxygen in a predictable manner without degradation by-products, and thus is termed by the developer *clean*. A transition metal catalyst introduces free radicals to trigger the scavenging process. A nonmigratory photoinitiator was developed so that the scavenging process could easily be started during the packaging step.

The Chevron Oxygen Scavenging Polymer (OSP) system begins with the two components. Comprising about 90% of the blend is the oxidizable ethylene methyl acrylate cyclohexane methyl acrylate (EMCM) resin. The second component, of 10%, is a concentrate that contains the proper ratio of photoinitiator plus a cobalt salt (transition metal catalyst).

A converter would incorporate these two resin components into a multilayer structure containing the OSP as an individual layer between a passive oxygen barrier, such as nylon, EVOH, or PET, and an inside sealant layer, such as LDPE, LLDPE, or ionomers The resin system can be engineered for conventional plastic processing conditions including cast and blown film, extrusion coating, blow molding, and others. At the time of packaging, the OSP layer is exposed to sufficient UV radiation to trigger the scavenging mechanism. Once the package is sealed, oxygen within the package would begin to be absorbed and fugitive oxygen would be absorbed as well. The induction time of the process can be varied based on the amount of component used.

UV activation can be tailored to meet specific application demands. The photoinitiator is reportedly responsive across a narrow UV range. Various UV power ranges, dosages, and wavelength ranges have been investigated with differing degrees of success.

The number of cyclohexene pendant groups attached to the polymer chain determines the scavenging capacity. The more pendant groups on the oxidizable polymer, the more oxygen that can be scavenged. The company has developed polymers with capacities ranging from 45 to 70 cc of O_2/gram of OSP. Demonstrations of oxygen scavenging rate have shown a given quantity of oxygen can be quickly removed from inside a package (Rodgers and Compton, 2000).

PALLADIUM CATALYSTS

The American Can Company palladium catalyst/hydrogen gas, oxygen scavenging technology cited earlier may have been derived from two earlier patents assigned to Universal Oil (Quesada and Neuzil, 1969). In these patents, the palladium catalyst was dispersed on alumina, i.e., porous aluminum oxide, and dried. Up to 0.5% palladium catalyst per weight of aluminum was employed. The catalyst may be within a sachet in the sealed food package or incorporated into the interior package walls in the ratio of up to 2 g per 15-cubic-inch container. The headspace gas is a blend containing up to 5% hydrogen and the balance nitrogen.

American Can Company's version was a flexible material named Miraflex 7™ (Zimmerman et al., 1974). The developers noted that the scavenging reaction produced water, which had to be trapped or removed since the contents were usually dry. Oxygen concentrations as low as 0.001% could be maintained over a six-month time frame.

A 1981 patent assigned to Ethyl Corporation (King, 1981) describes almost the same palladium catalyst as did the American Can Company King reported incorporation into the liner of a bottle or jar closure. An interesting aspect of this development is the incorporation of the finely divided catalyst to be blended into oxygen-permeable polyolefins or waxes. This layer is separated from the package interior by an oxygen-permeable ionomer layer. Above the reactants in the closure is a desiccant layer.

ENZYMES

Enzymes are biologically specific catalysts that accelerate biochemical reactions. For in-package use, the enzyme may be incorporated into the package structure. To function within a package material, the enzyme must be immobilized and the substrate, reactant, or a constituent must be circulated past the site to effect a reaction. Immobilization of an enzyme or placing it in a static position where it may function for an indefinite period may be accomplished by making the enzyme an integral part of the package material.

Although a broad range of enzymatic reactions arising from their incorporation into package materials might be conceived, only a relatively small number have actually been attempted on a practical basis, the major examples being oxygen removed by means of glucose oxidase plus catalase.

Enzymes and their actions have been known for many decades, but the notion of incorporating them into package materials to achieve a desirable result dates back only to the late 1940s. Almost simultaneously with the idea of protecting against phenolic browning of dry foods by removing residual oxygen, the notion of in-package glucose oxidase/catalase reactions was born. The initial action of the enzyme glucose oxidase is with residual quantities of glucose, a reducing sugar active in the nonenzymatic, nonoxidative Maillard browning reactions. Since highly reactive hydrogen peroxide is produced by glucose oxidase, it must be removed by catalase, which breaks it into water and oxygen. This concept was reduced to practice employing porous packets of the enzyme mix, in which the enzymes slowly reacted with minute quantities of residual oxygen.

The exponential growth of modified-atmosphere packaging in the 1980s led to the concepts of oxygen, carbon dioxide, and moisture control using in-package sachets of chemicals, including some enzymatic agents. During the 1980s, interest in this field increased with enzymologist Dr. John Budny, who proposed and, in some instances, physically evaluated glucose oxidase/catalase to remove oxygen. He tested immobilized enzymes within the interiors of package structures to catalyze reactions of products contained within packages (Budny, 1989 and 1990; Brody and Budny, 1995). The reaction is:

$$2G + 2O_2 + 2H_2O \rightarrow 2GO + 2H_2O_2$$

where G is the substrate.

Glucose oxidase transfers two hydrogens from the $-CHOH$ group of glucose to oxygen with the formation of glucono-delta-lactone and hydrogen peroxide. The lactone then reacts spontaneously with water to form gluconic acid. One mole of glucose consumes one mole of oxygen. To reach zero oxygen in a package with 500 cc headspace needs 0.0043 mole of glucose. The major variables are the speed of the enzyme reaction, the quantity of glucose available, and the rate at which oxygen enters into the package. In the presence of catalase, the hydrogen peroxide is broken down. One mole of glucose reacts with one-half mole of oxygen, thus decreasing the overall effectiveness of the system.

Since hydrogen peroxide is a very good oxidizing agent, it is "just as objectionable, or even more so, than is the original molecular oxygen" (Baker, 1949). Thus, catalase is introduced to break down the hydrogen peroxide:

$$2H_2O_2 + catalase \rightarrow 2H_2O + O_2 + catalase$$

The sum of these two reactions yields half the oxygen originally present, and therefore ultimately the free oxygen approaches zero.

This concept of enzyme incorporation into a package material was first overtly described in a 1956 patent (Sarett and Scott, 1956). In this patent, the enzymes glucose oxidase and catalase in a solution were impregnated into or onto a moisture-proof paper or fabric sheet. The enzyme was bound to the sheet with a water-dispersible adhesive such as polyvinyl alcohol, starch, casein, or carboxymethyl cellulose. The enzyme coating face must contact the moist product to ensure that the requisite oxygen-reduction reactions take place. The enzyme system was claimed to serve as a barrier to oxygen, which would otherwise have been transmitted through the sheet. Products described as benefiting from this system of oxygen reduction include cheese, butter, and frozen foods subject to phenolic browning. The package material is described as having ". . . an exposed surface covered with a gas-permeable packaging material and having an inter layer between and in contact with packaging material and . . . food . . . inter layer providing an oxygen barrier . . ." (Sarett and Scott, 1956). The specific package materials identified were moisture-proof cellophane, paper, and rubber hydrochloride, with impregnation employed for the papers and coating for the plastic and cellulose films. Also cited as being suitable substrates were wax paper, styrene, polyethylene, and vinyls. Oxidation of cheese surfaces was retarded by the presence of the enzyme-containing package material.

Scott, of Fermco Laboratories, published a paper on "Enzymatic Oxygen Removal from Packaged Foods" (Scott, 1958), which described enzymes that were incorporated into packaging materials or introduced into packets. Fermco Laboratories is a manufacturer of enzymes, one category of which was labeled Fermcozyme antioxidants, and of the packets that were named "Oxyban." This paper appeared to mark the first publication on the use of packets of oxygen-scavenging chemicals in packages.

Among the applications suggested in the publication were:

- aqueous foods in packets in situations in which the enzyme and the product should be kept separate
- nonaqueous foods in packets, as, for example, chow mein noodles

The dried enzyme system was coated on the surfaces of package materials for processed cheese. Deposition of the enzymes was in solution form or incorporation in a dry starch mixture prior to "dusting" the package material surface. When the dry and therefore inactive enzyme picked up moisture from the product, it was activated and was a sufficient oxygen interceptor to control formation of brown ring. Another series of experiments focused on obviating oxidative gray coloration on the surfaces of luncheon meats.

The "Oxyban" product was a dry glucose oxidase/catalase/glucose/buffers blend to be placed in small packets in which it reacted with oxygen in packages of roasted and ground coffee, smoked yeast, or egg solids.

Later, Scott elaborated on the oxygen-scavenging packet for in-package deoxygenation (Scott and Hammer, 1961). Using the same glucose oxidase/catalase packet system described earlier from their laboratory experiments, they proceeded to a more commercially viable mechanism. Among the issues they enumerated were oxygen scavenger surface area; the need for moisture activation; necessity to neutralize product, gluconic acid; avoidance of enzyme deactivation; and package material structure to pass oxygen but not moisture.

The gluconic acid problem was obviated with phosphate buffers. Fifteen grams of "Oxyban" enzyme mix in a packet was reported capable of removing all measurable oxygen from a sealed No. 2 size can at ambient temperature.

An interesting side note was an exploration of the use of glucose oxidase alone, which led to an increase in the amount of hydrogen peroxide, which, in turn, slowed the subsequent rate of oxygen uptake. The products benefited by the total system were primarily dry milk, potato granules, and ice cream mix.

A patent application from Finland (Lehtonen et al., 1991) described a package material containing an enzyme system to remove oxygen from the interior of the package by enzymatic reaction. The enzyme, e.g , glucose oxidase, was incorporated into a package material with a gas-impermeable layer on the exterior and a gas-permeable layer on the interior. The layer containing the oxygen-consuming enzyme was sandwiched between two plastic film layers.

The background of this patent cited a 1969 German publication describing the use of glucose oxidase in package materials for the surface protection of meats, fish, and cheese products. The classical active-packaging review paper by Labuza and Breene (1989) described a similar technology of coating plastic film with glucose oxidase catalase, with the enzyme system activated by moisture from the food.

The Lehtonen patent application detailed a flexible package structure containing an enzyme system in a liquid phase trapped between films, the outer of which might be polyamide (nylon) or polyvinylidene chloride–coated polyester. The inner film would be polyethylene, which generally is not a good gas barrier. The enzymes of choice were oxidases of the oxidoreductase family, using oxygenases and hydroxylases, which bind oxygen to oxidizable molecules. The enzyme solution contained a buffer and a stabilizer.

The films produced by application to substrates were employed either as the lidding film layer or as the thermoformable bottom layer for tray-type packages. As the temperature increases, the gas permeability of package materials increases, which also increases the ability of the enzyme system to reduce the oxygen content from the 20.9% in air to about 1% at ambient temperature within one day.

Budny (1989) added another dimension to packaging by allowing an individual package to become a processing unit, i.e., to become active. With the enzyme present, the package can perform a function that was previously limited to in-plant operations. Budny developed a two-enzyme system involving glucose oxidase and catalase to intercept oxygen and applied the technology for enzymes in active packaging to improve the concept of oxygen removal. In the development, liquid packaged in a polyethylene coextrusion, coated-paperboard, gable-top carton reacted enzymatically with glucose in the package wall to form gluconate. The resulting hydrogen peroxide was enzymatically reacted with catalase to produce oxygen and water that re-enter the contained product liquid. Budny described a container with an internal reactor, which is really an integral section of the package wall containing the enzymes through which the liquid contents flow and permit enzymes to react.

Another patent (Ernst, 1991) described a glucose/glucose oxidase enzyme mixture in a porous precipitated silica acid carrier. Calcium carbonate, calcium hydrogen phosphate, magnesium carbonate, or disodium hydrogen carbonate may also be employed as carriers or reaction accelerators. The oxygen scavenger may be in the interior of in-package sachets.

Labuza and Breene (1989) analyzed the issues of incorporating glucose oxidase into package materials. They suggested that to counteract the quantity of oxygen passing through a cracked or pinholed aluminum foil lamination, an enzyme surface will have to react with oxygen in the following manner:

Rate of oxygen passage $(cc/m^2/t)$ = permeability $\times$ area
$\times$ oxygen pressure difference between the outside and inside

$$Rate = 0.1 \times 1 \, [0.21 - 0.01] = 0.02 \, cc/m^2/day$$
$$= 20 \, \mu L/day$$

This calculation assumes air outside and less than 1% oxygen inside. For the worst case, i.e., with a pinhole or cracked score, there would be the need to scavenge 1 cc/day. A polymeric film could be made equivalent to an aluminum foil barrier by binding the oxygen-scavenging enzyme to the inside surface of the film to react with the excess oxygen.

Since many foods may have minimal contact with the package surface, except on the sides and bottom, this may not be the best approach for oxygen scavenging.

At 30 to 40°C, pure glucose oxidase has a rate of oxygen consumption of about 150,000 $\mu L/hr/mg$. Spreading the enzyme at a concentration of 1 mg per m^2 on a film would be equivalent to reacting all the oxygen coming through a film having a very high oxygen permeability of about 18,000 $cc/day/m^2$. Thus, a 1 m^2 surface with 1 mg of enzyme spread out on it should be able to consume all the oxygen entering through any package film. One advantage is that both polypropylene and polyethylene are good substrates for immobilizing enzymes. Another factor is the stability of the enzyme when bound to the film.

YEAST

At least two patents from the 1980s and 1990s describe the use of yeast to remove oxygen from the headspace of hermetically sealed packages. One patent, from enzyme manufacturer Gist Brocades, focused on the incorporation of immobilized yeast into the liner of a bottle closure (Edens et al., 1992). The other patent used the yeast in a pouch within the package (Nezat, 1985). The concept of the patents was that, when moistened, the yeast is activated and respires, consuming oxygen and producing carbon dioxide plus alcohol. In the bottle-closure application, any carbon dioxide and alcohol produced would enter the contents, in this case beer, without causing measurable changes in the product.

THE PACKAGE MATERIAL AS A REDUCING AGENT

An interesting concept from Japan's Nippon Steel involves using the package itself, in this instance a metal can, as the oxygen scavenger (Maeda et al., 1989). The inventors found that by providing a hydrophilic coating layer on the surface of a can body material, oxygen scavenging proceeds rapidly without oxidation of the base material. A hydrophilic coating and an oxygen-permeable layer are applied to function as an oxygen scavenging system. The material may be any metal reactive with oxygen, such as iron, zinc, or manganese, or materials other than metals, such as a plastic film with a thin metallic layer (0.5 to 20 mg per dm^2) of the metal deposited either by a vacuum deposition method or applied with an adhesive.

FERROUS IRON-BASED SCAVENGERS

The most significant oxygen scavengers in commercial use are those based on ferrous iron. These types of oxygen scavengers were commercially developed by Japan's Mitsubishi Gas Chemical Company under the tradename Ageless®, but are also available from other suppliers under other tradenames.

The product is basically ferrous iron oxide, which oxidizes to the ferric state. The water activity of the food must be high enough to provide moisture for the iron to oxidize, or another means or activation must be employed. A level of 0.01% (100 ppm) oxygen within closed packages is achievable.

The ferrous iron powder is separated from the food by keeping it in a small, highly oxygen-permeable sachet clearly labeled "Do not eat" because of the possible toxicity if the contents of the sachet were accidentally consumed. The LD_{50} (lethal dose that kills 50% of a rat population) for ferrous iron is 16 g/kg body weight. The largest sachet commercially available contains 7 g of ferrous iron, which would amount to only 0.1 g/kg for a 70-kg person, or 160 times less than the lethal dose for adults. The product has been approved by the Japanese Ministry of Health and the United States Food and Drug Association (FDA), provided there is a warning label on the package and that all the films and printing inks used conform to regulatory requirements. The U.S. Department of Agriculture (USDA) also approved its use in indirect contact for packaging beef jerky, dehydrated meat, and poultry products.

Under *ideal* stoichiometric conditions, 1 g of ferrous iron consumes 3.36 × 10^5 liters of oxygen if the iron is in solution and totally available, making this by far the most effective oxygen-scavenger material in commercial use. In the large particulate form, less iron is available, but in the ferrous iron powder form, the powder disintegrates rapidly and so most of the iron becomes available. For an 0.003-inch-gauge polyester film with an oxygen transmission rate of 4 cc/m^2/day, a sachet containing 1 g of ferrous iron should be able to handle the oxygen permeating in for 84 days for a 1-m^2 surface area package. This computation assumes that the oxidation rate to the ferric state is faster than the diffusion rate of oxygen into the pouch. Instead of a sachet, this amount of iron might be spread out over the entire surface area of a film if it were entrapped so as to not enter the food, as has been done by Multisorb Technologies in its Fresh Max® label product (Idol, 1993a) and by Mitsubishi Gas Chemical Company.

Assuming hermetic seals, the amount of ferrous iron that must be used depends on the initial oxygen level in the headspace, the amount of dissolved oxygen in the food or beverage, and the film oxygen-permeation rate. The general rule is that 1 g of ferrous iron reacts with 300 cc of oxygen. One Ageless® sachet configuration designed for contained product with water activities of less than 0.85 requires one to four days to reach 100 ppm residual headspace oxygen. In-package oxygen scavenger sachets are available in sizes that can react with 20 to 2,000 cc of oxygen, based on using packages with oxygen permeabilities no greater than 20 cc/m^2/day.

Other Ageless® types function better at higher water activities and have faster reaction rates (zero to two days) with the same oxygen scavenging capacity. One type does not absorb oxygen until exposed to water activity greater than 0.85 and so can be handled if maintained dry. Other types require that they be handled in a low or zero oxygen atmosphere since they begin to react immediately.

Various suppliers have demonstrated significant improvement in shelf life with the oxygen-scavenging sachet. For example, white bread shows mold growth at ambient temperature in air in four to five days, whereas with the sachet in the package, mold-free shelf life can be more than 45 days. Pizza crust, which molds in two to three days in air at 30°C (86°F), can be mold-free for more than ten days. Removal of oxygen also eliminates beetles in stored flour packages, which may be a significant problem in humid environments. Ageless® sachets are claimed to be able to keep packages of snack foods at less than 0.1% oxygen over a three-month period at ambient temperature.

It must be pointed out, however, that an oxygen-free atmosphere in foods with greater than 0.92 water activity can be conducive to the growth of pathogenic anaerobic microorganisms such as *Clostridium botulinum*. Thus, use of the sachet under certain conditions could be dangerous if the temperature is not kept at 3°C (36°F) or below. Several researchers have demonstrated that unprocessed fish packages containing sachets showed rapid development of botulism toxin in the fish without necessarily exhibiting signs of spoilage.

Another problem is that the iron-containing sachet can trigger on-line metal detectors, which is one rationale for using ascorbic acid in-package oxygen scavenger sachets.

AMERICAN ORGANIZATIONS, TWENTIETH CENTURY

Multisorb Technologies

Until the 1990s, most of the U.S. development of iron-based oxygen scavengers was from Multisorb Technologies, Inc. (formerly Multiform Desiccants, Inc.). Most of their early program was with sachets, FreshPax™, and more recently, with flat, attachable, adhesive-backed labels or patches tradenamed FreshMax®. Multisorb Technologies has been manufacturing oxygen scavenger sachets since 1988 (Idol, 1991, 1993a, 1993b; Powers, 1999). Their scavengers, all ferrous iron salt–based, have capacities ranging from 20 to 2,000 cc of oxygen. Among their commercial applications are bakery goods, processed meats, nuts, real bacon bits, chilled pasta, pizza crusts, fresh red meat, and also military rations (e.g., MRE baked goods and snack items).

Patents for Multiform Desiccants' commercial oxygen scavenger products are based on a combination of 20% to 25% ferrous iron plus 10% to 14% sodium chloride on silicon dioxide, enclosed in highly gas-permeable DuPont Tyvek® spun-bonded polypropylene sachet material (Cullen and Vaylen, 1991). The reasoning is that during formulation and handling, the scavenger is dry and therefore inactive. The silica water-attracting component has greater affinity for the electrolyte than does the oxygen-attracting component, and so the electrolyte does not combine with the iron. When exposed to a relative humidity of 90%, the moisture activates the electrolyte so that oxygen is scavenged by the iron.

A later Multiform Desiccants' patent (Cullen and Incorvia, 1992) described the same oxygen scavenging system for liquid contents, in which the dry components are in a cylindrical cartridge floating on the liquid within a can or jar.

In 1993, Multiform Desiccants was granted a patent on an oxygen absorber capable of removing oxygen from a low-moisture product (Cullen et al., 1993). In this development, the iron and the electrolyte are complemented by a hydrogel to provide water for activation. The hydrogel is an amorphous silicon dioxide that can contain up to 70% moisture and remain as a flowable powder. The hydrogel was reported to provide enough water to produce an electrolytic solution to cause a rapid reaction of the iron with oxygen. Products with moisture contents as low as those of potato chips were cited as examples of packaged products in which virtually 100% oxygen absorption is accomplished within 36 hours.

An interesting variation on the basic reduced iron system was disclosed in another Multiform Desiccants' patent (McKedy, 1993). The iron was particulate size annealed electrolytically reduced iron 100 to 325 mesh. According to the developers, electrolytically reduced iron (Fe°) absorbs oxygen at a faster rate than ferrous iron (Fe^{++}), especially at temperatures below 10°C. Annealing is reported to change the structure of the electrolytically reduced iron by increasing the surface area, which, in turn, causes the iron to be more active in oxygen absorbing capacity.

The Pactiv ActiveTech system for removing oxygen from fresh red meat distribution packaging involves affixing an oxygen scavenger-containing sachet (Tyvek® material) from Multisorb Technologies and simultaneously activating injection of an aqueous fluid (DelDuca, 2000).

Air Liquide

U.S. Patent 4,840,280 to Air Liquide (Schvester, 1989) is specifically directed to the use of ferrous iron to reduce oxygen in beverage packages such as beer bottles by incorporating the oxygen scavenger into the closure. To avoid the direct contact of the chemical with the liquid, a film that is very highly permeable to gaseous oxygen (10,000 cc/m^2/atm/hour or greater) and water vapor, but impermeable to liquids, is used as barrier between the scavenger and the contained liquid. The combination of oxygen scavenger and protective film is introduced during the closure manufacture. Once in place, the oxygen-absorption reaction is initiated and the residual oxygen in the container's headspace is removed. The patent indicates that the scavenger is iron oxide powder.

AMERICAN ORGANIZATIONS, TWENTY-FIRST CENTURY

BP Amoco

During the 1990s, Amoco Chemicals developed a ferrous-based package material adjunct that was later incorporated by Cadillac Products Packaging Company into an internal oxygen-scavenging film for military rations. The company name was changed to BP Amoco, and during this period they developed nonferrous ethylenic oxygen scavengers that could be linked to polyester backbones. In 2000, BP Amoco changed its name again to BP Chemical.

CIBA

During 2000, Ciba Specialty Chemicals Corporation acquired the ferrous Amosorb 2000 technology from BP Amoco, and is now marketing it as CIBA™ Shelfplus™ inorganic-based oxygen absorbers for polyethylene and polypropylene. Both companies' products are discussed in more detail in Chapter 4.

JAPANESE ORGANIZATIONS

Only a select few of the many Japanese companies offering oxygen scavengers are discussed here.

Mitsubishi Gas Chemical Company

Mitsubishi Gas Chemical Company is the world and United States leader in oxygen-scavenging technology and sales, especially for iron-based absorbers, although the company also manufactures ascorbic acid-based and other absorber products. They have been producing and marketing oxygen scavengers since 1977. All of their Ageless® oxygen scavengers are manufactured in Japan, and they have been marketed in the United States since 1984.

Mitsubishi Gas Chemical's iron technology is basically reduced iron powder (Fe^{++}) plus sodium chloride and zeolite in a gas-permeable sachet for placement in a hermetically sealed gas-barrier package or in a label to be affixed to the interior of a package

As early as 1979 (Komatsu et al., 1979), Mitsubishi Gas Chemical was granted a U.S. patent on sachets containing iron sulfite as the oxygen scavenger. The constituents of the oxygen absorber include iron salt, 0.1% sulfur to prevent evolution of hydrogen gas; 0.01 to 5 parts of a metal halide or sodium sulfate, sodium sulfite, or sodium bicarbonate; and inert materials. These materials are blended into a granular form and placed in perforated and gas-permeable polyethylene or cotton fabric pouches for placement inside hermetically sealed packages. When exposed to moisture, the iron compound reacts with oxygen. In the presence of fresh foods of relatively high water activity, no water is needed since the system absorbs the moisture from the internal environment.

Iron-based oxygen scavengers were applied into bottle closures in a 1981 Mitsubishi Gas Chemical Company patent (Moriya, 1981). In this application, a gas-permeable film pouch containing the oxygen scavengers was placed in the liner of the bottle closure and separated from the liquid contents by a gas-permeable membrane. The film prevented the oxygen scavenger from direct contact with the package contents.

In 1983, Mitsubishi Gas Chemical was granted a patent in which the oxygen scavenger in the gas-permeable pouch (sachet) was specifically applied to extend the shelf life of processed fish and roe packaged in gas-impermeable bags. Among the oxygen reactants mentioned in addition to reduced iron were iron carbide, hydroquinone, BHA, ascorbic acid, and erythorbic acid (Nakamura and Uchida, 1983).

Two 1984 Mitsubishi Gas Chemical patents (Ohtsuka et al., 1984) detailed the construction of the gas-permeable sachets to contain the oxygen absorber. Specifically cited were the gas permeabilities required for passage of air.

Among the materials suggested were spun-bonded polyolefins such as DuPont's Tyvek® spun-bonded polyolefin and microporous polypropylene such as Celanese Celguard®.

In 1987, Mitsubishi Gas Chemical Company received a patent for a cup-shaped "packet" that is effectively moisture-impermeable (Wakamatsu et al., 1987). In this development, the oxygen-scavenger packet contains the moisture required for activation The cup is closed with a gas-permeable membrane that is heat- or adhesive-sealed to the flange. This development is evidently intended to be a packet to be inserted into a jar of, for example, vitamin C tablets.

Another bottle closure oxygen scavenging device was described in 1988 (Morita et al., 1988). Again, a sachet containing the scavenging material is placed in a compartment in the liner of the closure separated from the liquid contents by a membrane. This development is characterized by the presence of a vacant compartment above or behind the scavenger sachet–containing compartment to assist in ensuring the hermetic seal of the closure

An important 1988 patent from Mitsubishi Gas Chemical Company (Inoue, 1988) describes a label sheet containing the oxygen scavenger In simple form, the label is a mixture of fibrous material, ferrous iron powder, water, and an electrolyte formed by a process similar to paper making. The developers recognized shortcomings that were, and continue to be, deterrents to commercial applications of oxygen scavengers·

- There is a risk that the consumer may accidentally consume the oxygen-scavenging agent.
- Since the sachet cannot be made into an extremely small size, it is not applicable to a package having a very small inner space.
- If a food container has a narrow opening, the oxygen scavenger cannot be inserted into the container.
- When an oxygen scavenger is to be fixed within a tray having a certain thickness, the oxygen scavenger causes an increase in the height of the tray, presenting challenges in manufacturing the tray and also in securely fixing the oxygen scavenger in place.
- Oxygen scavenging powder in a bag tends to coagulate into a lumpy shape and has a reduced surface area for contacting air. Therefore, to produce the scavenging of oxygen at a desired rate, it is necessary to provide a larger quantity of oxygen-scavenging powder than with the theoretical oxygen-scavenging capacity.

The 1988 development was intended to provide a sheet-like oxygen scavenger that can be securely fixed to the inner wall of a package.

A 1989 Mitsubishi Gas Chemical Company patent (Inoue, 1989) describes a sachet for oxygen scavengers specifically intended for use within retort pouches. The differentiating variable is the impregnation of the sachet material with a fluorine resin to coat the entire fiber to provide air, water, and fat resistance. The fluorine resins cited are various tetrafluorides and hexafluorides such as ethylene tetrafluoride.

Toyo Seikan

Among the largest package suppliers in the world, Toyo Seikan's marketing has been largely confined to the Far East. A 1993 patent (Koyama et al., 1993) issued to Toyo Seikan describes a film lamination, i.e., a package material containing a ferrous oxygen scavenger. Although the emphasis is on the incorporation of the oxygen scavenger into a bottle-closure liner, an alternative is a package film.

In conventional practice, since the resin has a high oxygen permeability and therefore admits outside air, the net oxygen-scavenging rate is much lower than the oxygen-scavenging rate attained when a paper-packaged oxygen-scavenging agent is directly charged, and the oxygen-absorbing effect is not satisfactory.

The primary objective of the development is to provide a heat-formable oxygen-absorbing resin. The development consists of an oxygen-absorbing resin, composed of a heat-formable resin, and an oxygen scavenger. The resin used is a blend of a polyolefin with a water-absorbing resin (such as a modified polyethylene oxide), a vinyl alcohol polymer, a sodium acrylate polymer, and an acrylic acid/vinyl alcohol copolymer with an olefin resin.

All of the oxygen scavengers customarily used were suggested for use singly or in mixtures of two or more. Further, these oxygen scavengers can be used in combination with accelerators, such as hydroxides, carbonates, sulfites, halides of alkali metals, and alkaline earth metals. The particle size of reduced iron ranged from 0.1 to 100 μm. When the particle size is smaller, the oxygen-absorbing capability is greater. However, if the particle size is smaller than 1.0 μm, heat is generated when mixing or kneading in air to blend, and so mixing in nitrogen becomes necessary.

As the oxidation promoter, a chloride of an alkali metal and/or alkaline earth metal such as NaCl plus $CaCl_2$ is preferred. If this oxidation promoter is used in combination with a manganese salt such as manganese chloride ($MnCl_2$), the absorption of oxygen is effectively enhanced. The reduced iron and oxidation promoter is used in a weight ratio ranging from 98%/2% to 90%/10%. The oxidation promoter could be 0.05% to 15.0% by weight of sodium chloride, 0.01% to 15.0% by weight of calcium chloride, and 0.01% to 5.0% by weight of manganese chloride.

Toyo Seikan was issued a patent on a multilayer retortable plastic can that incorporated an oxygen scavenger into a moisture-absorbing layer (Koyama et al., 1993). The basic oxygen barrier cited in the patent is ethylene vinyl alcohol (EVOH), whose water sensitivity has been widely publicized. Among the mechanics to obviate the loss of oxygen barrier as a result of water has been placement of a desiccant or water absorber adjacent to the EVOH. In this manner, water that penetrates the plastic at higher rates during retorting is absorbed before reaching the EVOH.

In this Toyo Seikan development, the oxygen scavenger is incorporated into the gas barrier, i.e., EVOH, which functions as a moisture absorber, particularly under retort conditions; the EVOH thus triggers the oxygen scavenger to remove oxygen. This concept is intriguing since the EVOH is a mediocre oxygen barrier when wet, and the oxygen scavenger only functions when wet. Thus, the oxygen scavenger, wet from the retorting process, compensates for the ineffectiveness of the EVOH during these circumstances. Evidently, the action of the oxygen scavenger must be triggered by the presence of a water-absorbing material such as sodium or calcium chloride.

The oxygen scavenger, essentially an iron compound, is incorporated into the moisture-absorbing oxygen-barrier resin layer sandwiched between the moisture-resistant inner and outer layers. The methods described for incorporation include melting EVOH copolymer resin and mixing it with the powdered iron particles at a ratio of 7% scavenger by weight. Results of testing of oxygen concentrations within trays thermoformed from the coextruded sheet indicate that the presence of the oxygen scavenger can reduce the headspace oxygen by factors of five to ten.

The concept of this structure marketed under the tradename Oxyguard™ appears to have merit as a means to provide an all-plastic structure in either flexible or semirigid form; Oxyguard™ could function as a significantly better oxygen barrier than a conventional material and, additionally, continuously remove oxygen from the package headspace. The technology has been reduced to developmental commercial practice in packages of aseptically packaged wet rice popular in Japan. These packages consist of thermoformed coextruded multilayered polypropylene/EVOH/polypropylene trays, conventionally with oxygen-scavenging sachets inserted. In Toyo Seikan's version, the scavenger is incorporated in the layer between the polypropylene and the EVOH.

Sumitomo Chemical

A 1992 patent (Kawakiti et al., 1992) assigned to Sumitomo Chemical describes an oxygen-absorbing plastic sheet characterized by small microvoids formed by stretching the plastic containing the oxygen absorber. Because the uniformly dispersed oxygen absorbent is in contact with air contained in and entering the microvoids, the stretched sheet effectively absorbs oxygen in the air. The oxygen absorbent is iron in the form of carbides or oxides. Suggested electrolytes include sodium, calcium, or magnesium chloride. The sheet was reported to not absorb significant amounts of oxygen at low relative humidities but to be activated when the relative humidity exceeded 50%.

This development is interesting in that monoaxial orientation produces microvoids where, according to the developers, the majority of the oxygen reaction occurs. Although dependent on the plastic to have a high oxygen permeability, the microvoid concept certainly increases the exposed surface area very significantly.

Toppan Printing

Japan's second most important producer and marketer of oxygen scavengers generally has emphasized ascorbic acid scavengers. In a 1983 patent, Toppan described a system in which iron salt is combined with ascorbic acid to produce the oxygen-scavenging effect, with the advantage that no moisture is required. In a typical reaction:

$$FeSO_4 \cdot 7H_2O + Na_2 SO_3 \cdot 7H_2O \rightarrow FeSO_3 + Na_2 SO_4 + 14 H_2O$$

The water supplied from the product and an alkali compound react with each other to absorb oxygen:

$$2FeSO_4 + 2 Ca(OH)_2 + H_2O + 1/2 O_2 \rightarrow 2Fe(OH)_3 + 2CaSO_4$$

These reactions are employed in concert with iron powder and ascorbic acid to remove oxygen.

Toppan uses a sachet of paper plus porous or perforated polyethylene film within the sealed package.

EUROPEAN ORGANIZATIONS

PLM

PLM developed an oxygen-scavenging system in which metal ions, presumably iron, are mixed in a plastic melt to be coextruded in film or sheet form (Anonymous, 1990b). As in so many other systems, the permeation of the oxygen through the plastic leads to a reaction. According to PLM, the system was also produced in a lacquer form to coat the interior of containers.

Atco SA

In Caen, France, Atco SA launched a range of oxygen absorbers in 1989 (Askenazi, 1989). Under the tradename Atco, these are ferrous iron powder in gas-permeable sachets intended to be placed in cheese packages (Legrand, 2000).

ASCORBIC ACID-BASED ABSORBERS

Second in commercial importance to iron-based compounds as oxygen scavengers are ascorbic acid and its derivatives. Because ascorbic acid is a six-carbon compound, relatively high weight is required to react with oxygen. Like iron, it is readily, but not quickly, oxidized to dehydroascorbic acid, which is relatively innocuous. Some argue that the reaction is not oxidation but rather hydrogen transfer; regardless, ascorbic acid reacts with and removes atmospheric oxygen The conversion to dehydroascorbic acid is catalyzed by the presence of metal ions such as iron.

According to Japan's Toppan Printing, iron compound scavengers do not function in the presence of carbon dioxide, whereas ascorbic acid functions well in the presence of carbon dioxide. Toppan employs their "C" or ascorbic acid-type for applications in the presence of carbon dioxide such as in packages of roasted and ground coffee and bread. Among the other attributes of ascorbic acid as an oxygen scavenger is its lack of response to metal detectors on packaging lines.

Grace's Daraform® oxygen scavengers incorporated into plastic bottle-closure linings have been either ascorbic acid analogues or ligands

Darex® Container Products (now a unit of Grace Performance Chemicals, Lexington, Massachusetts) has announced an ethylene vinyl alcohol with proprietary oxygen scavenger developed in conjunction with Kuraray Co., Ltd. (Okayama, Japan). Grace is an intermediate product company (i.e., they buy raw pellets and liquid-based plastisols). They compound and test wet and dry oxygen scavengers in single-layer extrusions for bottle-cap gaskets and internal can liners for rigid packaging applications, e.g., to extend the shelf life of products such as beer and baby food. Their oxygen-scavenger products are available in the United States, Europe, and Japan. In dry form, pellets containing unsaturated hydrocarbon polymers with a cobalt catalyst are employed as oxygen scavengers in mechanical closures, plastic and metal caps, and steel crowns (both polyvinyl chloride [PVC] and non-PVC lined). They reportedly can extend the shelf life of beer by 25%. Grace also formulates liquid-based plastisols (PVC based) containing an ascorbic acid–sulfite mix for oxygen scavengers (the catalyst is copper sulfate) in PVC plasticizers, for large caps.

Pillsbury

Dr. Ernst Graf was granted two patents (Graf, 1994b) assigned to the Pillsbury Company, covering the use of ascorbic acid analogues whose action could be accelerated by the presence of copper, a reaction generally avoided in most food situations. Dr Graf recognized that copper is highly reactive in food, i.e., it accelerates lipid oxidation. The developer indicated that the oxygen removal with copper plus ascorbic acid is so rapid, i.e., 2- to 3 minutes, that no oxygen could be detected in a closed package. This development is not for packaging but rather for incorporation of the scavenger directly into the food. It was used by Pillsbury for several refrigerated Mexican food sauces marketed during the late 1980s.

W. R. Grace

W. R. Grace has been one of the few U.S. companies directly involved in the development of oxygen-scavenging package materials. W. R. Grace's oxygen-absorbent bottle-gasket liners have been marketed in both the United States and Europe since about 1990. The original patent (Hofeldt and White, 1992) describes a sealing compound formed from a composition that can be extruded into a bottle closure to become a gasket. The developers noted that the compound can be either a bottle liner or a liner in a double-seamed can end. The material, essentially ascorbic acid in a plastisol compound, can resist thermal pasteurization, but it is not retortable if that food process happens to be an objective.

In a patent directed to beer packages (Hofeldt and White, 1993a), the developers indicate that the same oxygen absorbers in the same polymeric matrix are activated by thermal pasteurization or sterilization. Although the relative humidity in the headspace of a beverage-containing bottle or can is generally sufficiently high to activate the scavenger, the reaction could be accelerated by heating to increase moisture permeation into the scavenger mix. Thus, they suggest a scavenger that remains effectively inert until pasteurization or sterilization. Moisture generated from the interior contents permeates into and becomes trapped within the gasket to accelerate the oxygen-removal reaction. Oxygen trapped in the headspace and that permeating into the bottle through the gasket are both reacted with the scavenger.

Japanese Organizations/Companies

At least three other Japanese patents are among the literature describing ascorbic acid analogues for oxygen scavenging. Two describe pouches containing ascorbic acid, alkaline compounds, an aluminum salt, a reaction accelerator, and silica gel (Ueno and Tabata, 1992 and 1993). The objective is to increase the flowability of the oxygen-absorbent powder in order to increase the speed of filling the pouches.

A Japanese patent from Toray (Yamada et al., 1992) describes the construction of a laminated flexible material label with an oxygen-permeable membrane containing an ascorbic acid oxygen scavenger. An oxygen-permeable film covers the oxygen absorber, which includes an asymmetric porous membrane whose outer surface is formed as a dense skin layer. Since the dense layer is very thin but porous, oxygen transmission to the oxygen absorber through the oxygen-permeable film is high. The oxygen scavenging chemical is within a nonwoven material trapped between the porous layer and a high barrier backing.

ECONOMICS

Questions exist concerning the economics of introducing iron powder into independent sachets or component compartments. Cost appears to be an obstacle to the commercial acceptance of in-package scavenger sachets. A sachet capable of removing 25 to 30 cc of oxygen would cost about U.S.$0.02 to U S.$0.03 To reduce oxygen to 0.3% without mechanical vacuum would require up to U.S.$0.10 to U.S.$0.15 of oxygen scavenging sachets. Although 1 g of iron can theoretically remove 200 cc of oxygen, the efficiency is about half that, due to particle agglomeration. With the wide experience of using iron as an oxygen scavenger, however, it should be expected that in all applications in which oxygen is a problem, this will be the first material proposed and tested.

DEVELOPMENTAL

One of the more provocative of the oxygen-scavenger products for package materials emerged in 1991 from U S. chewing gum manufacturer Wm. Wrigley, Jr. The objective of the Wrigley development reported in patent form (Cartwright et al., 1992) is to provide a long-term flavor-delivery system through emission from materials in the package structure. During the course of this development, the researchers postulated that the reverse reaction, oxygen scavenging, could be performed by the same technology. The proposed technology suggests incorporating the oxygen scavenger into porous plastic beads that, in turn, are embedded as a coating in the primary package material immediately adjacent to the product. Although they suggested that iron be employed as the scavenger, virtually any oxygen scavenger could be used. The researchers used Mitsubishi Gas Chemical's Ageless® iron oxide/activated charcoal as the oxygen scavenger.

The Wrigley development involves mixing the iron/activated charcoal with edible oil and soaking the porous beads with the mixture. The loaded microbeads are then blended with a mixture of microcrystalline and paraffin waxes that are then used as the laminant in a polyvinylidene chloride (PVdC) (exterior)/polyethylene lamination. Oxygen from within the package passes through the polyethylene into the oxygen scavenger layer.

The Wrigley research was based on porous plastic beads developed by Advanced Polymer Systems of Redwood City, California, capable of adsorbing and desorbing gaseous compounds. In a joint venture with Dow Corning, the product was marketed by the larger partner under the tradename Polytrap.

OXBAR™

Carnaud MetalBox's Oxbar™ system is an integration of gas-barrier packaging and oxygen scavenging and was suggested for packaging oxygen-sensitive products (Cochran et al , 1991). Oxbar™ is a chemical barrier system developed specifically to exclude oxygen and possesses the following attributes:

- It is oxygen specific, i.e , there is no removal of other components
- It is polymer based
- It provides a total barrier to oxygen.
- It works independent of temperature and humidity.
- It has a long, but finite, lifetime.

Developmental focus was on the polyester (PET)/nylon MXD6/cobalt system because of the extensive research done on this combination. The reactive components are the MXD6 nylon (polymetaxylylene adipamide or polymetaxylylene diamine-hexanoic acid) and a cobalt salt that catalyzes the reaction of MXD6 with the oxygen passing through the package wall.

Oxbar™ is a polymeric oxygen scavenging system and is therefore readily formed into containers using standard thermoplastic converting techniques, such as injection and extrusion, injection blow molding. It has been reported that the materials may also be formed into films.

The low levels of reactive ingredients (MXD6 and cobalt) reportedly do not significantly modify the physical or processing characteristics of the structural polymer but impart a gray color. The materials are combined together as a blend to offer a monolithic structure on existing PET bottle-converting equipment.

The Oxbar™ system is reported to provide virtually total barrier to oxygen ingress for as long as the reactants are present. For example, in 1-liter PET-only versus PET/MXD6 (4%)/Co-blend bottles, the reference PET-only bottles show an oxygen transmission value of 3 5 $cc/m^2/day$ at 22°C (73°F), whereas the equivalent Oxbar™ variable has a transmission value of less than 0.04 $cc/m^2/day$. The very low level of transmission persists for a period approaching two years under these conditions.

An important feature of the Oxbar™ system is its ability to maintain its barrier properties under elevated temperature and relative humidity conditions. Many barrier polymers exhibit performance loss under conditions of high temperature or high relative humidity.

In common with other barrier systems, Oxbar™ relies on chemical activity. Thus its barrier properties persist for only a finite time, which is a function of two primary variables. scavenging capacity and rate of consumption. Scavenging capacity depends on the composition of the blend (MXD6 and cobalt content) and the container wall thickness. Even after the "zero" transmission period is ended, there is very slow return over several months to the expected blend value. Thus, even for low-capacity Oxbar™ blends, there is a significant reduction in cumulative oxygen ingress into the package.

Carnaud MetalBox reported on tests conducted with bottles containing various oxygen-sensitive beverages including beer, orange juice, and wine. Shelf lives of over six months were demonstrated. In these bottles, enough oxygen scavenging capacity was present to consume more than one year's oxygen entering through polyester bottles. Essentially no oxygen passed into the package in one year.

By itself, nylon MXD6 has FDA approval in the United States for use as a blend with PET at levels of up to 30%. The temperature of use should not exceed 49°C and the alcohol content of the product should not exceed 8%. These conditions cover a range of uses such as cold-filled beer and carbonated beverages.

At their commercial introduction, cobalt salts of organic acids had not been used as polymer additives Extractability tests using a range of stimulants and conditions detected no cobalt migration from Oxbar™ containers. Nevertheless, because of the paucity of data on the end-products of nylon MXD6 oxidation, the company did not apply to the FDA for Oxbar™ use for packaging aqueous products.

The system could be used in any low-oxygen-permeability plastic with which it is compatible. The limit in oxygen permeability for the effective structural plastic is in the range of five times that of PET, but not much more. The system has been tried in polypropylene (P = 150) (Adams et al., 1991) in a multilayer structure including polyester skin and the nylon MXD6 oxygen-absorber system. The developers reported that the introduction of the oxygen-scavenger system reduced the oxygen permeation a substantial two orders of magnitude but cited only a factor of 60 to 300. The final oxygen permeation was reported to be less than 2 cc O_2/m^2/atm/day.

Continental PET Technologies in the United States has received patents for multilayer transparent bottles containing a blend of polyester and nylon MXD6 and of polyester, nylon MXD6, and cobalt (Collette, 1991) The nylon MXD6-containing oxygen-scavenging system is present only in the core layer of the multilayer bottle, which is coinjected with conventional bottle-grade polyester in order to reduce the overall bottle percentage of nylon MXD6 to less than 1% by weight. The MXD6 constitutes 5 + 2% of the core layer blend, which, in turn, constitutes about 4% of the total injection-molded preform or bottle. At this level, the developers reported clarity in both injection-molded preforms and in final bottles blown from the preforms.

The system is the same as or very close to that reportedly used by Continental PET Technologies for Miller beer bottles with the PET/MXD6/Co in a core layer between two PET layers in an injection-stretch blow-molded bottle (Brody, 1999)

The current FDA attitude is toward permitting approval for materials with potential migration problems. Thus, if the blend is separated from the interior contents by at least 0.001-inch of virgin barrier plastic, this structure should not encounter regulatory questions. This interpretation, if valid, might not be applicable for film formation because of the thinner gauges.

AQUANAUTICS/ADVANCED OXYGEN TECHNOLOGIES (AOT)/ W. R. GRACE

Advanced Oxygen Technologies, Inc., formerly Aquanautics Corporation, developed well-publicized scavenger products as a series of oxygen-binding complexes for applications that include the removal of trace residual oxygen from beverages. These compounds, organometallic ligands, function in a fashion similar to that of hemoglobin, which transports oxygen in blood; however, the products have been reported to be simpler, synthetically produced molecules. A number of these complexes have been immobilized on insoluble supports in functional package material form. These immobilized oxygen binders have been demonstrated to be effective at reducing oxygen levels from aqueous solutions, including beer, to as low as 10 ppb. They can be used as part of the package itself, such as the closure, to remove oxygen from the product through the headspace. In addition, the company claims they can incorporate the materials into plastic film and sheet form.

Proof of the scientific principle of ligands has been demonstrated in the form of oxygen binders immobilized on silica and other supports. These reductions to practice resulted in the engineering and production of commercial oxygen-absorbing systems for application in the food and beverage industries. The term "ligands" refers to small molecules used for oxygen binding and absorption. Target applications to date appear to be liners for closures for beer, wine, and fruit juice bottles and other oxygen-sensitive foods and beverages.

PHOTOSENSITIVE DYES

Australian researchers from Commonwealth Scientific and Industrial Research Organization (CSIRO, now Food Science Australia) in North Ryde, New South Wales, Australia, have published extensively on chemical systems for oxygen scavenging. One of their systems used photosensitive dyes, among other compounds. They have suggested that when oxygen is excited to the singlet state, the reaction is much faster than with iron compounds.

One process utilizes reactions inside the packaging film containing a photosensitizing dye and an excited oxygen trap. On illumination of the exterior of the film with visible light, dye molecules become excited and pass their excitation to oxygen as it diffuses into the film from either the package headspace or from the liquid food. The excited oxygen molecules react with the trap and become bound, i.e., are scavenged. While the film is illuminated, the process continues until all the oxygen is reacted.

The reaction scheme is:

$$dye + light \rightarrow dye*$$

$$dye* + O_2 \rightarrow *O_2$$

$$*O_2 + trap \rightarrow bound\ oxygen$$

where * represents an excited state of the species. Polyketones can act as photosensitizers.

This photochemical process offers advantages over most of the other oxygen-removal processes because it requires no addition of sachets to the food package, does not involve particulate matter in the transparent packaging material, and can be reversed with light. However, scavenging does not occur in the dark and so a good oxygen-barrier plastic in place of aluminum foil would be necessary for highly sensitive foods.

The polymer system has been improved to bind the reaction products as part of the package to prevent migration into the food. A small pouch lined with a scavenger film area of 150 cc and containing 25 cc of air can be deoxygenated in about 15 minutes.

In this CSIRO process, a photosensitive dye is impregnated onto ethyl cellulose film. On exposure to ultraviolet light, the dye activates oxygen to the singlet state. This oxygen then can react with any acceptor to form an oxide The dye first used was erythrosine, an FDA-approved food-color additive plus a color sensitizer that is bleached by light. The singlet-oxygen acceptors tested were difurylidene erythritol (DFE), tetraphenyl prophine (TPP), dioctyl phthalate (DOT), and dimethyl anthracene (DMA), but these are not FDA approved for food contact.

A 1994 international patent application from CSIRO describes oxygen scavengers that function independent of transition metal catalysts (Rooney, 1993). The system contains a reducible organic compound that is reduced under predetermined conditions. The reduced form of the compound is oxidizable by molecular oxygen so that its reduction and/or subsequent oxidation occurs independently of the presence of a transition metal catalyst. The reduction and/or subsequent oxidation of the one reducible organic component is also independent of the presence of an alkali or acid catalyst. The reducible organic compound may be reduced under predetermined conditions such as by exposure to light, heat, gamma or electron beam irradiation, or corona discharge.

Typically, the reducible organic compound has the capacity to be converted to an excited state such as a triplet form. The triplet form then becomes reduced to an essentially stable state by gaining or abstracting an electron or hydrogen atom from other molecules, or by redistributing an electron or hydrogen atom within the compound itself. The reduced molecule is reactive toward molecular oxygen to produce activated species such as hydrogen peroxide, a hydroperoxy radical, or a superoxide ion (Rooney, 2000) Preferably, the reducible organic compound should be stable in air at room temperature or should be in its fully oxidized state. Examples of such compounds include substituted anthraquinone, such as 2-methylanthraquinone, which can absorb in the ultraviolet spectrum. The reducible organic compound component may constitute up to 50% by weight of the material.

Co-monomers functioning as the polymer backbone may be any ethylenically monounsaturated or polyunsaturated substance such as vinyl acetate, vinyl alcohol, acrylonitrile, methacrylonitrile, and acrylic monomers

SILGAN CONTAINERS CORPORATION

Polystar Packaging, Inc., and successor company Silgan Containers Corporation, have performed considerable research work in the area of oxygen scavengers. In the area of inorganic scavengers, the following systems have been tested: Zn/Pd, Sn/FeSn$_6$, Fe/alkaline metals, and/or earth alkaline metals. Ox/Redox metal and non-metal salts, Sn^{++}, Sn^{++++}, Fe^{++}/Fe^{+++}, SO$_3^{--}$/SO$_4^{--}$, etc. In the area of organic oxygen scavengers, the following have been tested. quinone/hydroquinone and dehydro-ascorbic acid, ascorbic acid, unsaturated polymers "Kratons"/Co^{++} and/or Mn^{++}, Pd^{++}, Cu^{++} catalysts, and the same with ultraviolet or thermal activation, addition of initiators "benzophenone and derivatives." Incorporation of any of the above systems in packaging, however, led to sequential problems during compounding; coextrusion; injection molding; filling and sealing; heat processing the filled container, storage of the processed container; and finally, heating the container in a microwave oven for consumption. Ultimately, Silgan's work resulted in the development of an all-plastic, shelf-stable, easy-open food container that can be heated in a microwave oven. Their iron-based oxygen-scavenging plastic lidding material is FDA-approved for direct food contact, is microwavable, and reportedly has a significant large capacity and fast absorption rate (Hekal, 2000).

BIBLIOGRAPHY

Adams, J. F, M. A. Cochran, R. Folland, J. W. Nicholas, and M. E R. Robinson 1991 "Packaging." U. S Patent 5,049,624. September 17

Anonymous. 1990a. "Oxbar: The Oxygen Barrier." *Food Engineering International*, 15(8):51.

Anonymous. 1990b. "PLM Develops New Barrier Technique for Active Packages" *Pack Marknaden*, (4):20

Askenazi, B 1989 "Oxygen-Absorbing Sachets." *Emballage Digest*, 337:72–74, 76

Baker, Dwight. 1949 "Deoxygenation process" U S Patent 2,482,724. September 20

Bloch, Felix. 1965. "Preservative of oxygen-labile substances, e.g., foods " U S Patent 3,169,068. February 9

Brody, A. L 1999. "Oxygen Scavenging Packaging: Where Are We? Where Do We Go From Here?" *Proceedings of "Oxygen Absorbers· 2000 and Beyond" Conference*, Chicago, Illinois, George O Schroeder Associates, Inc , Appleton, Wisconsin. June

Brody, A L. and J A Budny. 1995. "Enzymes as Active Packaging Agents." In *Active Food Packaging*, M L. Rooney (ed.). Glasgow, UK: Blackie Academic & Professional, London. pp 174–192

Budny, John. 1989. "A transporting storage or dispensing container with enzymatic reactor." International Patent Application, WO89/06273.

Budny, John. 1990. Presentation at *Pack Alimentaire*. San Francisco, California, May.

Cartwright, W R , C P DiGrigoli, and L G. Kasbo. 1992. "Flavor-release material " Canadian Patent 2,044,078 April 23.

Cochran, Michael A , Rickworth Folland, James W Nicholas, and Melvin E R. Robinson. 1991 "Packaging " U S. Patent 5,021,515 June 4

Collette, Wayne. 1991. "Recyclable multilayer plastic preform and container blown therefrom." U.S Patent 5,077,111. December 31

Cook, P. 1999. "Stealth Scavenging Systems: A Remarkable Breakthrough Technology" *Proceedings of "Oxygen Absorbers: 2000 and Beyond" Conference*, Chicago, Illinois, George O. Schroeder Associates, Inc , Appleton, Wisconsin June.

Cullen, John and Samuel Incorvia. 1992. "Flexible oxygen-absorbing cartridge." U S. Patent 5,092,914. March 3.

Cullen, John S. and Nicholas E. Vaylen 1991 "Oxygen-absorbing package, composition and method of formulation thereof." U.S Patent 4,992,410. February 12.

Cullen, John S , George E McKedy, Christopher S. Nigon, and Thomas H. Powers. 1993. "Oxygen absorber for low moisture products." U.S. Patent 5,207,943. May 4

DelDuca, Gary. 2000 "Innovative Packaging Concepts for Meat and Poultry" Institute of Food Technologists' Annual Meeting Presentation. June.

Edens, Luppo, F Farin, A F Ligtvoet, and J Van Der Platt. 1992. "Dry yeast immobilized in wax or paraffin for scavenging oxygen " U.S. Patent 5,106,633. April 21

Ernst, R 1991 "Oxygen absorbent and use thereof" U S Patent 5,028,578. July 2.

Farrell, Christopher J , and Boh C Tsai 1987. "Oxygen scavenger" U.S. Patent 4,702,966 October 27.

Graf, Ernst 1994. "Oxygen removal " U.S. Patent 5,284,871. February 8.

Hekal, Ihab 2000. Silgan Containers Corporation. Norwalk, Connecticut.: Personal communication

Hofeldt, R and S. White. 1992 "Sealed containers and sealing compositions for them." U.S. Patent 5,106,886 April 21.

Hofeldt, R. and S. White. 1993a. "Sealed containers and sealing compositions for them " U S Patent 5,204,389. April 20.

Hofeldt, R and S. White. 1993b. "Sealed containers and sealing compositions for them " U S Patent 5,227,411. July 13.

Idol, R. C. 1991. "A Critical Review of In-Package Oxygen Scavengers." *Sixth International Conference on Controlled/Modified Atmosphere/Vacuum Packaging*, Schotland Business Research, Inc., Princeton, New Jersey.

Idol, R. C. 1993a. "A Retail Application for Oxygen Absorbers in Europe." *Pack Alimentaire '93*, Schotland Business Research, Inc., Princeton, New Jersey.

Idol, R. C. 1993b. "Oxygen Absorbing Labels." *Packaging Week*, 9:16.

Inoue, Y. 1988. "Sheet-like, oxygen-scavenging agent." U.S. Patent 4,769,175. September 6.

Inoue, Y. 1989. "Oxygen absorbent package" U.S Patent 4,856,650. August 15.

Inoue, Y. and T. Komatsu. 1990. "Oxygen absorbent." U.S. Patent 4,908,151. March 13.

Inoue, Yoshiaki, S. Murabayashi, K. Fujiname, and I. Yoshino. 1994. "Oxygen absorbent composition and method of preserving article with the same." U.S. Patent 5,286,407. February 15.

Kawakiti, T., T. Kume, K. Nakag, and M. Sugiyama. 1992. "Oxygen absorbing sheet: molded, stretched low density ethylene copolymer containing iron powder with electrolyte on surface." U.S. Patent 5,089,323 February 18.

King, Roderick. 1981. "Closure with oxygen scavenging system." U.S. Patent 4,279,350. July 21.

Komatsu, Toshio, Y. Inoue, and M. Yuyama. 1979. "Oxygen absorbent." U S. Patent 4,166,807. September 4

Koyama, Masayasu, Y. Oda, and M. Yamada. 1993. "Plastic multilayer vessel" U.S. Patent 5,153,038. October 6.

Labuza, T P. and W. M. Breene. 1989. "Applications of Active Packaging for Improvement of Shelf-Life and Nutritional Quality of Fresh and Extended Shelf-Life Foods." *Icelandic Conference on Nutritional Impact of Food Processing*, Reykjavik, Iceland, 1987; *J. Food Processing and Preservation*, 13:1–69.

Legrand, Marc. 2000. "Active Packaging Technologies in the European Food Industry." *Proceedings, International Conference on Active and Intelligent Packaging*. Campden & Chorleywood Food Research Association, U.K., September.

Lehtonen, Paavo, Pirkko Aal Tonen, Ulla Karilainen, Risto Jaakkola, and Seppo Kymalainen 1991. "A packaging material which removes oxygen from a package and a method of producing the material." International Patent Application WO 91/13556. September 19.

Loo, Ching C. and William P. Jackson. 1958. Carnation Company. "In-package oxygen Remover." U.S. Patent 2,825,651. March 4.

Maeda, Shigeyoshi, Tsungtoshi Asai, Hidejiro Asano, and Haruyoshi Taguchi. 1989. "Materials having a deoxidation function and a method of removing oxygen in sealed containers." U S. Patent 4,877,664. October 31.

McKedy, George E. 1993. "Oxygen absorber." U S. Patent 5,262,375. November 16.

Miltz, J., P. Hoojjat, J. K. Ham, J. R. Giacin, B. R. Harte, and I. J. Gray. 1988. "Food Packaging Interactions." J. H. Hotchkiss (ed.). *American Chemical Society Symposium Series*, 365:33.

Morita, Yoshikazu, T. Komatsu, and Y. Inoue. 1988. "Oxygen scavenger container used for cap." U.S Patent 4,756,436. July 12.

Moriya, Takehiko. 1981. "Container sealing member with oxygen absorbent." U.S. Patent 4,287,994. September 8.

Nakamura, Hisao and Youji Uchida 1983 "Novel method of storing processed fish and roe." U.S. Patent 4,399,161. August 16

Nezat, Jerry W. 1985. "Composition for absorbing oxygen and carrier thereof." U.S. Patent 4,510,162 April 9.

Ohtsuka, Sadao, T. Komatsu, Y. Kondoh, and H. Takahashi 1984 "Oxygen absorbent packaging." U.S. Patent 4,485,133. November 27.

Powers, T. 1999 "'Designing In' an Oxygen Absorber." *Proceedings of "Oxygen Absorbers: 2000 and Beyond" Conference*. George O. Schroeder Associates, Inc. Appleton, Wisconsin. June.

Quesada, Camilo and Richard Neuzil. 1969. "Oxygen scavenging from closed containers" U.S. Patent 3,437,428. April 8.

Rho, K L , P. A Seib, O K. Chung, and D. S. Chung. 1986. "Retardation of Rancidity in Deep Fried Instant Noodles (Ramyon)." *Journal of American Oil Chemists Society*, 63(2):251.

Rodgers, B. D and L. Compton. 2001. "New Polymeric Oxygen Scavenging System for Coextruded Packaging Structures" *Proceedings of "Oxygen Absorbers: 2001 and Beyond" Conference* George O Schroeder Associates, Inc. Appleton, Wisconsin. June

Rooney, M. L 1984. "Photosensitive Oxygen Scavenger Films: An Alternative to Vacuum Packaging" *CSIRO Food Research*, 43:9–11 March.

Rooney, M L 1993 "Oxygen scavengers independent of transition metal catalysts." International Patent Application PCT/AU 93/00598. November 24.

Rooney, Michael L. 2000. "Application of ZERO$_2$™ Oxygen Scavenging Films for Food and Beverage Products." *Proceedings, International Conference on Active and Intelligent Packaging*, Campden & Chorleywood Food Research Association, U K. September.

Sarett, Ben L and Don Scott. 1956. "Enzyme treated sheet product and article wrapped therewith" U S Patent 2765233.

Scholle, William R. 1977. "Multiple wall packaging material containing sulfite compound." U.S. Patent 4,041,209. August 9

Schvester, Pascal 1989 "Sealing cap for liquid food or beverage containers" U.S Patent 4,840,280. June 20.

Scott, Don. 1958 "Enzymatic Oxygen Removal from Packaged Foods" *Food Technology*, 12(7):7

Scott, Don and Frank Hammer. 1961. "Oxygen Scavenging Packet for In-Packet Deoxygenation." *Food Technology*, 15 12

Speer, Drew V. and William P. Roberts. 1994 "Improved oxygen scavenging compositions for low temperature use" International Patent Application WO 94/07379. April 14

Speer, Drew V., William P. Roberts, and Charles R Morgan 1993. "Methods and compositions for oxygen scavenging" U S. Patent 5,211,875 May 18.

Sugihara, Y., T. Kashiba, H. Hatakeyama, and T. Takeuchi. 1993. "Oxygen absorbent." U S. Patent 5,180,518. January 19.

Sugihara, Yasuo, T. Takeuchi, H Wakabayashi, A. Hosomi, and T. Komatsu 1992. "Oxygen absorbent" U.S Patent 5,102,673. April 7.

Throp, Arnold. 1980. "Closures for liquid product containers." U.S. Patent 4,188,457. February 12.

Ueno, Ryuzo and A. Tabata. 1992. "Oxygen absorbent." U.S Patent 5,128,060. July 7

Ueno, Ryuzo and Akihiko Tabata 1993 "Oxygen absorbent." U.S. Patent 5,236,617 August 17

Valyi, Emery. 1977. "Composite materials" U.S. Patent 4,048,361. September 13.

Wakamatsu, Syuji, T. Komatsu, Y. Inoue, and Y. Harima. 1987. "Oxygen absorbent packet." U.S. Patent 4,667,814. May 26

Yamada, Shinichi, Isamu Sakuma, Yoshio Himeshima, Takao Aoki, Tadahiro Uemura, and Akira Shirakura. 1992. "Oxygen scavenger" U.S. Patent 5,143,763 September 1.

Yoshikawa, Y, A Ameniya, T. Komatsu, and T. Inone. 1977. "Oxygen Absorbers" *Japan Kokai*, 104:4, 486.

Zimmerman, P. L, L J Ernst, and W F Ossian. 1974 "Scavenger Pouch Protects Oxygen Sensitive Foods" *Food Technology*, 28:8, 103

CHAPTER 4

Oxygen Scavenging in 2000 and Beyond

ONLY in the last years of the twentieth century did several lines of research on oxygen-scavenging plastics appear to be coming to fruition. Previously, the potential benefits of such active oxygen-removing plastics had been appreciated, but apparently simpler approaches to maintaining oxidative stability of foods were found to be more acceptable technically and economically. As discussed earlier, iron-based sachets have been and continue to be the in-package oxygen-scavenging technology of commercial choice. Although limited by reason of requiring water for activation, color, and limited rate and capacity, they have been the simplest and have had the fewest secondary effects. Most of the applications have been reduced iron powder in high-gas-transmission, flexible sachets inserted into the high-oxygen-barrier food package. About ten years ago, the iron powder was incorporated into flat coupons or labels to obviate the potential for a consumer to eat the iron powder.

LABELS

Iron-filled plastic strips have been developed by Japan's Mitsubishi Gas Chemical Company, Inc., with their Ageless® iron oxide materials. Sachets of iron powder were originally used hot-melt-bonded to lidding film in tray packs to reduce the probability of consumers eating them. It was potentially possible for the powder to be spilled if the sachet was cut with a knife when opening the package This potential problem has now been eliminated by dispersion of the powder in microporous plastic strips held in metallized self-adhesive labels These labels are claimed to render the package microwavable due to the extent of separation of the metal particles Porous labels are not as effective as those that contain the iron powder in free-flowing state because less iron is available for reaction with oxygen (Sakakibara, 2000).

As indicated earlier in this text, Multisorb Technologies, Inc., formerly Multisorb Desiccants, Inc., developed an analogous label called FreshMax® that is affixed to the interior of processed-meat packages.

INCORPORATION OF OXYGEN SCAVENGERS INTO PLASTIC STRUCTURES

Attempts to incorporate oxygen-scavenging chemistry into plastic packaging film structures have not yet resulted in corresponding commercial success, even though some oxygen-scavenging packages have been commercially introduced. Japan's Toyo Seikan and Mitsubishi Gas Chemical have produced trays with ferrous iron dispersed in sandwiched layers, but without substantial market penetration thus far. The difficulty experienced in making rusting reactions occur in a hydrophobic polymer matrix is that the sealant polymer is normally relatively impermeable to water vapor at ambient temperatures. Therefore, such systems are best applied to thermally stabilized packages in which water from the food contents is present in generous quantities for prolonged periods.

Several factors that impact on successful research outcomes in the field of oxygen-scavenging plastics are the following:

- effects of scavenging chemistry on physical properties of the plastic material employed to make the package structure
- access of the co-reactant, such as water, to the scavenging reagents in the plastic
- limitation of scavenging rate by oxygen permeability of the polymer
- appearance if solid particles are dispersed in the plastic
- premature reaction if the chemistry is not triggerable by an external action
- migration of any low-molecular-weight components, by-products, or reaction end-products into the product through the inner plastic

Although several plastics have been formulated by blending low-molecular-weight reagents into polymers, the most attractive alternative to dispersal of metallic or ferrous iron in plastics is the organic reactions of the polymers themselves. It appears that the rapid reaction rate achieved with high-surface-area ferrous iron powder in sachets is likely to be matched in plastics only by very rapid reactions, provided that the oxygen permeability of the plastic is not the limiting variable.

Table 7 gives an indication of the diversity behind a representative selection of some of the most recent developments.

TABLE 7. Oxygen Scavenger Blending in Plastics.

Mechanism	Activation	Company	Tradename	Comments
Polymer oxidation	Light, etc.	CSIRO/Southcorp Packaging	ZERO$_2$®	Under development
	Compounding	CarnaudMetalbox (Crown Cork & Seal)	OxBar™	Under development
	Compounding, metal, and light	Continental PET Technologies	n/a	Nylon MXD6 plus cobalt catalyst
	Time/ultraviolet light	Cryovac-Sealed Air	OS1000	Poly(1,2,-butadiene)
	Metal salt	Chevron Chemical		Benzyl acrylate copolymers
	Metal salt	BP Amoco P.L.C.	Amosorb® 3000	Polybutadiene
	Water vapor	Ciba Specialty Chemicals (formerly BP Amoco)	Shelfplus O$_2$ 2400 + 2500 Amosorb® 2000	Metal/salt
	Metal salts	Toppan Printing (Japan)	n/a	Polyolefins
Metal complex oxidation	Water vapor	Cryovac	Smart Mix	Advanced Oxygen Technologies patents

SANDWICHING OXYGEN SCAVENGERS

U.S. regulatory (FDA) approval for sandwiching of post-consumer-recycled polyester (PCR PET) between two layers of virgin PET in soft drink bottles evidently has facilitated the introduction of oxygen scavenging reagents and other oxygen scavenging technologies between layers of polyester plastic in packages such as beer bottles. Beer is highly oxygen sensitive; therefore almost complete removal of oxygen from the product and package interior is required.

The PET bottle does not present a serious problem with premature reaction due to its relatively low oxygen permeability. Continental PET Technologies apparently believed that the original CarnaudMetalbox Oxbar™ nylon MXD6/cobalt salt technology required too much time to become active, and the use of materials such as 2% blends of polyketones in PET to produce a plasticizing effect of water is required to enhance the oxidation reaction. Technologies for thin film applications, however, need an additional feature to prevent premature reaction if they are to provide maximum scavenging capacity

The light-activated, transition-metal-catalyzed process of Cryovac Sealed Air Corporation approached this by preplanned activation involving a slow generation of full capacity by consumption of antioxidants. This type of film, involving side-chain oxidation of a polydiene, ethylenically unsaturated hydrocarbon or other oxidizable polymer with glass transition temperatures well below 0°C, appeared to be designed as a permeation barrier for chilled products

In 1998, Cryovac announced its light-activated polymeric oxygen scavenging film, Cryovac® OS1000. According to Cryovac, other oxygen-scavenging systems require excessive heat and/or moisture for activation They developed a polymeric oxygen-scavenging (OS) system that absorbs oxygen in modified atmosphere packages and also serves as a total oxygen barrier when activated. This OS film is transparent and moisture independent. The objective of their research was to demonstrate the benefits of a polymeric oxygen-scavenging system in a variety of food products. Their results indicated that, when used as a headspace scavenger, the activated OS film reduced oxygen levels in modified atmosphere packages from 0.5% to 1.0%, to less than 500 ppm within seven days. In processed meats, improved color retention in comparison with conventional modified atmosphere packages was demonstrated. In shelf-stable processed tomato products packaged in flexible pouches, the rate of color degradation was reduced by 50% over 18 months. The polymeric OS film proved to be effective in delaying surface yeast and mold growth in fresh pasta when compared to conventional modified atmosphere packaging. It also delayed the onset of aerobic microbial growth in fresh modified atmosphere-packaged pasta. Significant advantages of this polymeric OS system were reported: extension of product shelf life; enablement of the production of transparent, flexible, or rigid packages of comparable appearance; and improvement of performance at costs comparable to existing package structures.

In modified atmosphere packaging applications, differences in packaging machines, packaging speeds, and vacuum/gas flush system efficiencies may result in variability in the initial residual oxygen levels of packages. Most often, off-line oxygen residuals are in the range of 2% or below. It is also possible for some products (e.g., pasta, bread) to outgas oxygen, increasing the residual oxygen and diluting the initial atmosphere within the first 24 hours.

OS1000 was a transparent, moisture-independent, polymer-based oxygen scavenging film. This material was a multilayer barrier lidstock for trays and incorporated a unique, proprietary oxygen scavenging layer coextruded in the film. OS1000 was to be used to produce oxygen scavenging packages on a variety of horizontal thermoform/fill/vacuum/gas flush/seal (HFFS) packaging machinery such as Multivac™ and Tiromat™. OS1000 was inactive until "triggered."

The proprietary oxygen scavenging layer consisted of three primary components: an oxidizable polyolefinic polymer, a photoinitiator, and a catalyst. The oxidizable polymer was the component responsible for binding the oxygen molecules. The photoinitiator absorbed the ultraviolet radiation and provided the energy to start the reaction. The catalyst helped to increase the rate of the scavenging reaction.

Triggering, the activation of the oxygen scavenging reaction, was accomplished by exposing the film to ultraviolet (UV-C) radiation. The Cryovac® Model 4100 Triggering Unit was used in conjunction with the packaging machine to actuate the oxygen scavenging reaction. Once the film was triggered, the oxygen scavenging reaction continued until either the headspace oxygen was consumed or the capacity of the film was exhausted

Cryovac has apparently recognized that the reaction of their oxygen scavenging polymer with oxygen can produce end-products that should be removed. Such end-products may be malodorous or toxic if they migrate through the inner plastic into the food contents. Cryovac has patented mechanisms to remove end-product compounds. Other participants such as Chevron Chemical have also patented analogous adsorbents whose purpose is to remove polymer oxidation end-products.

In 2000, Cryovac announced the termination of their OS1000 oxygen scavenging film project. They have elected to join with Chevron Chemical using the latter's oxygen scavenger in Cryovac film

Continental PET Technologies has taken a different approach in their use of polyketones as oxygen scavengers. The benefit of polyketones is claimed to be enhanced levels of carbon dioxide generated in the walls of containers such as beverage bottles. Commercial development has taken place in the PET bottle industry for use in beer bottling. The process is activated by hot compounding with water absorbed from the product and by ultraviolet light. Continental

PET Technologies has declined to publicly disclose its oxygen-scavenging chemistry, but patents and other information imply they are employing nylon MXD6 with transition metal catalyst in a polyester substrate sandwiched between two virgin polyester layers in the bottle wall and base.

RESIN-BASED SYSTEMS

Amoco Chemicals offered a resin-based system, Amosorb® 2000, which carried water-activated, oxygen-absorbing components. In their patents, Amoco Chemicals describes the use of iron in the presence of electrolytes and pH modifier in polymer films. The composition of Amosorb® 2000 has not been revealed, however, being described only as consisting of substances generally recognized as safe (GRAS) in the United States. The absorbent layer will be separated from the food by a sealant layer at least 12.5-mm (0.0005-inch) thick in plastic film structures and 25-mm (0.001-inch) thick in multi-layer laminate sheets. In June 2000, the Amosorb® 2000 oxygen absorber technology was acquired by Ciba Specialty Chemicals Corporation. The tradename was changed to Ciba™ Shelfplus™ O2-2400 (polyethylene carrier resin for blown film) (Table 8) and Shelfplus™ O2-2500 (polypropylene carrier for retort packages) (Table 9).

TABLE 8. Ciba Specialty Chemicals Shelfplus™ O_2-2400 Oxygen Absorber Specifications.

Characterization	Shelfplus™ O_2-2400 is an oxygen absorber used in food and beverage packaging to absorb oxygen
Chemical name	Proprietary
Applications	Shelfplus™ O_2-2400 is used in food and beverage packaging to absorb oxygen This absorber can be incorporated into a range of package structures, including the side walls or lids of rigid containers, flexible films, and closure liners The resulting package structure not only absorbs the oxygen initially present in the headspace and entrapped in the food or beverage, but also absorbs oxygen that permeates into the package over time
Features/benefits	Shelfplus™ O_2-2400 is an active barrier material with a finite amount of oxygen-absorption capacity
	Shelfplus™ O_2-2400 can be incorporated easily into multi-layer food-packaging structures without major changes in the current package design or fabrication equipment
	Using Shelfplus™ O_2-2400 can increase the shelf life of the product and help maintain product freshness
	Shelfplus™ O_2-2400 is thermally stable up to 260°C and can be processed at normal polyethylene-processing temperature of 200 to 250°C

TABLE 8. (continued).

	Shelfplus™ O$_2$-2400 is moisture activated so the packaged foodstuffs require water-activity levels that are 0 7 or higher in order to initiate the absorption of oxygen
Product forms	Code: Shelfplus™ O$_2$-2400 Appearance: free-flowing pellet
Guidelines for use	Shelfplus™ O$_2$-2400 is incorporated into a polyethylene layer of a multilayer structure To be most effective, a passive layer should be used in conjunction with the Shelfplus™ O$_2$-2400 layer to prevent the oxygen outside the container from consuming the absorption capacity of the package
	The Shelfplus™ O$_2$-2400 layer should not be the food contact layer to avoid potential interactions with foodstuffs that could affect the taste of the product
	To calculate the amount of Shelfplus™ O$_2$-2400 to be used, first determine the amount of oxygen to be absorbed Package storage temperature, water activity of contents, and the permeability of the package influence the overall capacity for oxygen absorption

Physical properties		
Density[1] (g/cc)	1 42	
Moisture[2] (wt%)	<0 05	
Melting point[3] (°C)	94	
Processing range (°C)	200–240	
Melt index[4] at 190°C (g/10 min)	3 0	
Melt flow rate[4] at 230°C (g/10 min)	—	
Carrier resin	Polyethylene (PE)	
Oxygen uptake capacity[5] (cc O$_2$/g Shelfplus™)	18+	
Target application	Blown film	

Registration	
	United States Food and Drug Administration (U S FDA)
	All components of Shelfplus™ O$_2$-2400 are determined to be GRAS (generally recognized as safe) for use in multilayer food packaging Use of Shelfplus™ O$_2$-2400 in multilayer construction is in compliance with the U S Federal Foods, Drug, and Cosmetic Act and all applicable food-additive regulations
	European Community Compliance Statement (EC)
	Shelfplus™ O$_2$-2400 is also in compliance with European Community regulations relating to plastics materials and articles intended to come into contact with foodstuffs
	United States Toxic Substances Control Act (U S TSCA)
	Shelfplus™ O$_2$-2400 is listed on the U S TSCA inventory The inventory status for various countries, in addition to the U S TSCA inventory, is included in the MSDS for Shelfplus™ O$_2$-2400

[1] ASTM D1505
[2] ASTM D4019
[3] ASTM D2117
[4] ASTM D1288
[5] Ciba test method

TABLE 9. Ciba Specialty Chemicals Shelfplus™ O$_2$-2500 Oxygen Absorber Specifications.

Characterization	Shelfplus™ O$_2$-2500 is an oxygen absorber used in food and beverage packaging to absorb oxygen
Chemical name	Proprietary
Applications	Shelfplus™ O$_2$-2500 is used in food and beverage packaging to absorb oxygen This absorber can be incorporated into a range of package structures, including the side walls or lids of rigid containers, flexible films, and closure liners The resulting package structure not only absorbs the oxygen initially present in the headspace and entrapped in the food or beverage, but also absorbs oxygen that permeates into the package over time
Features/benefits	Shelfplus™ O$_2$-2500 is an active barrier material with a finite amount of oxygen-absorption capacity Shelfplus™ O$_2$-2500 can be incorporated easily into multi-layer food-packaging structures without major changes in the current package design or fabrication equipment Using Shelfplus™ O$_2$-2500 can increase the shelf life of the product and help maintain product freshness Shelfplus™ O$_2$-2500 is thermally stable up to 260°C and can be processed at normal polypropylene-processing temperature of 200 to 250°C Shelfplus™ O$_2$-2500 is moisture activated so the packaged foodstuffs require water activity levels that are 0 7 or higher in order to initiate the absorption of oxygen
Product forms	Code: Shelfplus™ O$_2$-2500 Appearance free-flowing pellet
Guidelines for use	Shelfplus™ O$_2$-2500 is incorporated into a polypropylene layer of a multilayer structure To be most effective, a passive layer should be used in conjunction with the Shelfplus™ O$_2$-2500 layer to prevent the oxygen outside the container from consuming the absorption capacity of the package The Shelfplus™ O$_2$-2500 layer should not be the food contact layer to avoid potential interactions with foodstuffs that could affect the taste of the product To calculate the amount of Shelfplus™ O$_2$-2500 to be used, first determine the amount of oxygen to be absorbed Package storage temperature, water activity of contents, and the permeability of the package influence the overall capacity for oxygen absorption
Physical properties	Density[1] (g/cc) — 1 44 Moisture[2] (wt%) — <0 05 Melting point[3] (°C) — 150 Processing range (°C) — 220–250 Melt index[4] at 190°C (g/10 min) — Melt flow rate[4] at 230°C (g/10 min) — 6 0 Carrier resin — Polypropylene (PP)

TABLE 9. (continued).

	Oxygen uptake capacity[5] (cc O_2/g Shelfplus)	12+
	Target application	Retort containers
Registration	*United States Food and Drug Administration (U S FDA)*	
	All components of Shelfplus™ O_2-2500 are determined to be GRAS (generally recognized as safe) for use in multilayer food packaging Use of Shelfplus™ O_2-2500 in multilayer construction is in full compliance with the U S Federal Foods, Drug, and Cosmetic Act and all applicable food-additive regulations	
	European Community Compliance Statement (EC)	
	Shelfplus™ O_2-2500 is also in compliance with European Community regulations relating to plastics materials and articles intended to come into contact with foodstuffs	
	United States Toxic Substances Control Act (U S TSCA)	
	Shelfplus™ O_2-2500 is listed on the U S TSCA inventory The inventory status for various countries, in addition to the U S TSCA inventory, is included in the MSDS for Shelfplus™ O_2 2500	

[1] ASTM D1505
[2] ASTM D4019
[3] ASTM D2117
[4] ASTM D1288
[5] Ciba test method

Amosorb® and Shelfplus™ O_2 are a family of polymer-based, oxygen-scavenging concentrate pellets supplied to plastic-processing customers in free-flowing pellet form. The pellets reportedly can be incorporated into a range of package structures, including the sidewall of rigid containers, flexible films, and jar closure liners. The resulting structure not only absorbs the oxygen initially present and/or generated from the inside of the package, but also retards oxygen permeation into the package. The oxygen scavenger-containing package can be processed using conventional equipment ranging from hot-fill to retort. Amosorb® 2000 was effective in absorbing oxygen over a wide range of application temperatures, from 4°C for refrigerated foods, to 22°C for shelf-stable foods, to 130°C for prevention of food oxidation during retorting.

Three series of Amosorb® were commercially available: Amosorb® 1000, Amosorb® 2000, and Amosorb® 3000 Amosorb® 1000 was used for retorted or high-heat-treated, semirigid food-packaging applications. Amosorb® 2000, now Ciba™ Shelfplus™ O_2-2400 and O_2-2500, is for retorted and nonretorted food and refrigerated food applications. Amosorb® 2000 replaced Amosorb® 1000 Amosorb® 3000 is a modified polyester (recyclable copolyester) that may be injection-stretch blow-molded for bottle and jar-packaging applications for oxygen-sensitive beverage and food products. This material (in container and closure liners) has been employed as the core layer in polyester beer bottles that were introduced commercially for a brief period during 1999, e g., Anheuser-Busch beer market test. Amosorb® 3000 is currently available from BP Chemical.

In the Amosorb® 2000 and Shelfplus™ O_2 series, various grades based on carrier resins such as polypropylene, polyethylene, and elastomers and on melt flow rates ranging from 1 to 10 (grams per 10 minutes), are commonly available.

Amosorb® and Shelfplus™ O_2 are intended to enhance the oxygen-barrier properties of multilayer plastic sheets and films used to package foods subject to condition of use as demanding as microwave and retort applications. The layer in which the oxygen scavenger is mixed is separated from the packaged food by a food-contact layer that is at least 0.0005-inch thick for film structures and 0.001-inch thick for multilaminate sheets.

Amosorb® 2000 is thermally stable at temperatures as high as 260°C (500°F) for extruding polyolefin, and Amosorb® 3000 at higher temperatures for extruding PET. The processing range of Shelfplus™ O_2-2400 is 200 to 240°C (390 to 464°F) and of Shelfplus™ O_2-2500, 220 to 250°C (428 to 482°F).

U.S. bottle closure liner maker Tri-Seal (a division of Tekni-Plex) offers a closure lining material engineered to absorb oxygen initially present in bottle or jar headspaces. The material, trademarked TriSO$_2$RB™, is a coextruded sandwich of a foam core impregnated with Amosorb® scavenger, with virgin polymer skins on each side. One version, based on polyethylene, is for non-hot-fill applications. Another version is based on polypropylene and is intended for hot-fill applications (Moschitto, 1999).

In test results, a two-layer crystallized polyester (CPET) tray containing Amosorb® was fabricated to demonstrate the effectiveness of Amosorb® in a rigid container structure. The volume was 500 cc, and the total sidewall thickness was 0.026 inch. The outer layer was 0.013 inch of virgin CPET, and the inner layer was 0.013 inch of a blend of Amosorb® and virgin CPET.

In an Amoco study, the Amosorb® 1000-containing tray and a control tray without Amosorb® were retorted at 100°C (212°F) for 60 minutes prior to the oxygen permeation test. A small amount of oxygen, 2%, was injected into the trays to simulate the headspace oxygen initially present in the inside of the tray. After one year, the headspace oxygen of the Amosorb®-containing tray (0.5% O_2) was not only much lower than that of the CPET control tray (approximately 3.5% O_2), but also significantly lower than the initial oxygen level of 2%.

Similar results were obtained by testing other Amosorb®-containing structures: a single-layer CPET tray and a three-layer CPET tray. In the single-layer tray, Amosorb® was uniformly dispersed in the sidewall. In the three-layer tray, Amosorb® was dispersed in the middle CPET regrind layer, which was sandwiched between two virgin CPET layers.

Since many film structures for food packaging are multilayer, Amosorb® or Shelfplus™ O_2 can be incorporated into one of the layers.

Amosorb® was also claimed to be effective at 4°C for refrigerated food applications. The oxygen-absorption rate of Amosorb® increases with increasing temperature. As foods become more susceptible to oxidation and require more protection from oxygen at higher temperatures, Amosorb® is claimed to provide the protection.

Since active barrier packaging materials have a finite amount of oxygen-absorption capacity, they require protection from oxygen on the outside of the package so that oxygen-absorption capacity is not wasted.

UNITED STATES MILITARY APPLICATIONS

OXYGEN SCAVENGING FILMS

Although ration components are often specifically formulated for extended shelf stability, food degradation over time is still a problem. Packaging systems with integral, active mechanisms to optimize ration preservation are being developed by the U.S. military.

The U.S. military has a desire to eliminate oxygen scavenger sachets in packaged rations while still reducing the level of oxygen inside the package. The U.S. Army's Natick Soldier Center is working with industry to incorporate oxygen scavengers into their conventional trilaminate pouch materials containing aluminum foil, in order to maintain food quality and provide the food components with the military's minimum required shelf life of three years at 27°C (80°F). The typical preformed pouch for bread, cakes, and brownies is constructed, from inside out, of 0.002-inch gauge ionomer or polyethylene film, 0.00035-inch gauge aluminum foil, and 0.0005-inch gauge polyester.

The ability of the oxygen scavenging package to equal or exceed the current oxygen scavenger sachet's capabilities is being demonstrated by examining the oxygen content in hermetically sealed packaged products and by conducting storage studies, sensory evaluations, and microbiological and chemical analyses (e.g., moisture and hexanal) to determine the acceptability of the packaged rations over time.

Oxygen-scavenging films offer several advantages over sachets. Benefits include potential use with retort items, elimination of food product distortion that may occur when a sachet contacts the food, and potential cost savings due to increased production efficiency (i.e., simplified packaging and filling process) and convenience. Use of these films to package current ration components will eliminate the need for oxygen scavenger sachets while ensuring maximum safety for the soldier and/or customer. Oxygen-scavenging films will enable a wider variety of packaged ration components, including shelf-stable baked, snack, and thermostabilized items.

BAKERY AND OTHER NON-RETORT APPLICATIONS

In 1995, a research and development contract to develop oxygen-scavenging films was awarded to Cadillac Products, Inc. (now Cadillac Products Packaging Company, Troy, Michigan). Cadillac Products is also a major producer of non-retort packaging for the MRE ration. An aluminum-foil-laminated, four-ply packaging film incorporating oxygen-scavenging additives resulted from the joint developmental effort between the Natick Soldier Center, Cadillac Products, Inc , and BP Amoco Chemicals (Naperville, Illinois). The packaging material contained BP Amoco Chemicals' Amosorb® ABPA-2000-series oxygen scavenger concentrate within the sealant layer of the pouch structure. Cadillac's resulting ABSO$_2$RB™ film was the first material of its kind to be produced in the United States. The successfully leveraged efforts resulted in three food-packaging awards, including a 1996 DuPont Award for Innovation in Food Packaging Technology. The collaborative effort also resulted in patent applications for various aspects of the technology. Natick Soldier Center continues to work with Ciba Specialty Chemicals and their Shelf-plus™ O$_2$ technology. Both BP Amoco and Ciba Specialty Chemicals report they are also working with other companies to develop packages for commercial applications.

Natick Soldier Center's initial objective under the research and development contract was to replace oxygen scavenger sachets with an oxygen-scavenging film. In order to meet the oxygen absorption rate requirements for MRE (Meal, Ready-to-Eat) bread, however, it was determined that gas flushing was additionally needed to reduce the initial level of headspace oxygen in the package. Storage studies were then designed and conducted that included several packaging variables, including bread pouches with Multisorb Technology's Fresh-Pax™ oxygen scavenger sachets; pouches with only ABSO$_2$RB™ film; pouches with ABSO$_2$RB™ film and 100% nitrogen flushing (to less than 5% oxygen); pouches that included a mold-inhibiting film from DuPont Packaging and Industrial Polymers (Wilmington, Delaware); and control pouches with 100% nitrogen flushing and no oxygen scavengers. The main concern with the latter variable was the ability to meet the three-year shelf life requirement, since the porosity of baked goods causes control of residual oxygen by gas flushing alone to be challenging.

Shelf-stable bread pouches containing the Amosorb ABPA-2000 and nitrogen flushing maintained zero headspace oxygen levels through 36 months of storage at 27°C (80°F). Studies of improved prototype pouches with nitrogen flushing were also completed. These pouches demonstrated a much faster rate of oxygen absorption and a significant improvement in physical appearance compared with previously tested prototypes. With the final package structure, the Natick Soldier Center will conduct a scale-up order of selected horizontal form/fill/seal (HFFS), oxygen-scavenging pouch prototypes.

Films incorporating oxygen scavengers may be used to package oxygen-sensitive MRE ration components such as peanut butter, cheese spread, and freeze-dried entrees, in which the addition of a separate sachet is not feasible. In the future, oxygen-scavenging materials may also be used to package newly developed, shelf-stable ration items such as eat-on-the-move pocket sandwiches, snacks, calzones, pizzas, burritos, fruit pies, and high-energy performance-enhancing bars.

RETORT APPLICATIONS

Natick Soldier Center has also partnered with industry to investigate the feasibility of oxygen scavenging film technology for thermally processed rations, which typically exhibit off-odors, off-flavors, and a loss of texture when stored for long periods of time. Efforts began with Lawson Mardon, Smurfit-Stone Container Corporation, BP Amoco Chemicals, and Ciba Specialty Chemicals to investigate the feasibility of using oxygen scavenging flexible films for MRE retort entrees and polymeric group-serving tray lidstock

The current preformed pouch for retort entrees is constructed, from inside out, of 0.003- to 0.004-inch gauge polyolefin, 0.00035- to 0 0007-inch gauge aluminum foil, and 0.005-inch gauge pigmented polyester. Alternatively, the retort pouch may be an HFFS pouch consisting of a formed, tray-shaped body with a flat-sheet, heat-sealable cover. The tray-shaped body is constructed, from inside out, of 0.003- to 0.004-inch gauge polyolefin, 0.0015- to 0.00175-inch gauge aluminum foil, and 0.0010- to 0.0014-inch gauge oriented polypropylene; the flat-sheet stock for the cover is the same trilamination construction used for the preformed retort pouch. Alternatively, nylon (0.0006-inch gauge) may be added to the laminations of both pouches to improve handling performance, especially in cold weather environments (to −30°C/−20°F). Current military retort pouch manufacturers are Smurfit-Stone Container Corporation (Schaumburg, Illinois) and Pechiney Performance Plastics (formerly American National Can Company, Neenah, Wisconsin). Foil laminations, commercially produced by Lawson Mardon Packaging (Switzerland and Germany), are cold-formed in the horizontal mode on the Multivac R530 MC92 Fully Automatic Rollstock Vacuum Packaging machine. Exxon is the major domestic supplier of the polypropylene for the preformed retort pouches. Potential suppliers of the nylon-reinforced multilaminate film for the polymeric tray-lidding applications include the suppliers mentioned above, as well as Fuji Tokushu Shigyo Company, Ltd., and Dai Nippon Printing Company, Ltd., both of Japan.

In addition, oxygen-absorbent, polypropylene/ethylene vinyl alcohol rigid trays for retort foods have been jointly developed by Rexam Containers (Union, Missouri) and BP Amoco Chemicals under Natick's Polymeric Tray Program, which was initiated in 1996. The objective of this program is (1) to design, verify, and field a polymeric tray for shelf-stable retort and bakery items, and (2) to develop and evaluate components and technologies, such as three-year shelf-life polymeric tray materials and high-strength lidding materials, to enhance the capabilities of existing and future ration systems. The program was initiated primarily due to the discovery of gray spots, i.e., areas of internal corrosion, in the military's tin-plated steel traycans. A Process Action Team formed in 1996 determined that the polymeric tray presented the best opportunity to meet the military's future heat-and-serve feeding needs. Additionally, oxygen-absorbent polymeric trays begin to control long-term oxidative degradation of thermostabilized food products during the retort process ($120°C/250°F$ for two hours or more).

The Rexam polypropylene/ethylene vinyl alcohol/polypropylene polymeric tray was modified to include an oxygen-scavenging concentrate, BP Amoco Chemicals' Amosorb® ABPA-2000, for extended shelf life capability; an impact copolymer (polypropylene) for improved cold weather performance; rounded tray corners for better equipment interface; filled radii for increased rigidity; and a fiberboard sleeve for increased tray and lid survivability. In addition, a quadlamination film lid incorporating an added nylon layer was demonstrated to be superior at both ambient and frozen ($-30°C/-20°F$) conditions when compared to the trilamination film previously tested. A paperboard insert capable of oven use has also been developed for bakery applications, with the product first baked in the insert and then hermetically sealed alongside oxygen scavengers within the non-oven-use tray. Preliminary studies of this bakery concept have shown excellent moisture and water-activity stability through storage, as well as rough handling survivability.

Field tests conducted to date in temperate and arctic climates have demonstrated the climatic suitability of the polymeric tray and compatibility with all current field food-service equipment.

Shelf-life studies were initiated in 1996 on two retort food items packed in standard polymeric trays, oxygen scavenging polymeric trays, and metal traycan controls. Sensory evaluation testing after storage (six weeks at $49°C/120°F$; 12 months at $38°C/100°F$; and 36 months at $27°C/80°F$) showed no significant differences between the three packaging variables. In addition to these studies conducted at ambient RH environmental conditions, storage studies at moderate and extreme temperatures and RH conditions were initiated in July 1998 on four retort food items packaged in standard and oxygen-scavenging polymeric trays. Current sensory data do not yet convincingly support selection of the oxygen-scavenging tray over the standard polymeric tray, since both trays continue to meet military shelf-life requirements. There appears to be a noted quality benefit from the oxygen-scavenging additive, however, as sensory scores for this variable have generally been slightly higher. Results from this latter test will validate the predicted three-year shelf-life capability of the standard and/or oxygen scavenging tray at all current military storage locations in the United States and overseas

A performance specification for the polymeric tray, allowing for the use of the oxygen scavenger, was written in early 1997 and updated in 1998 based on the results of producibility studies and in-house test results. The polymeric tray was approved for limited procurement on October 15, 1998, and based on in-house storage studies and the results from large-scale user field exercises in the summer and fall of 1999, it was approved for full-scale procurement. Although the Polymeric Tray Program officially ended in September 1998, refinement efforts are continuing under the Field Group Ration Improvement Program. In October 1999, a Polymeric Tray Optimization Program was initiated, with an emphasis on developing the next generation of polymeric tray (Trottier, 2000).

By enhancing the quality and acceptability of packaged ration components, oxygen scavenging packaging materials may provide novel solutions to the problem of food degradation over time, as well as to ration underconsumption. Comprehensive programs by current U.S developers are most likely the most efficient route to deliver commercial package structures for military ration-packaging applications (Kline, 1999).

PREMATURE REACTION TO OXYGEN

The issue of premature reaction has been addressed in a radically different way in a system that involves converting the plastic resins into the package material while the scavenger is in a stable oxidized state. A patent application by Rooney and his colleagues at Food Science Australia describes methods of rendering their UV-triggered scavenger polymers further activatable by water. Research has been and is being conducted in collaboration with Southcorp Packaging (Sydney, Australia). The ZERO$_2$® oxygen-scavenging plastics developed so far reportedly have been tested successfully at temperatures from freezer to retort conditions, either alone or as blends in the sealant layer, as a buried layer in multilayer packages, or as the adhesive, i.e., laminating agent. The nomenclature "ZERO$_2$" describes an ingredient integrated into the polymer backbones of package material polymers such as polyethylene, polypropylene, and PET.

A limitation on any structure in which the reactive component is not in intimate contact with the package headspace is the time required for oxygen to diffuse to the reaction centers. This is particularly the case for chilled or frozen foods, i.e., foods at below ambient temperatures. Not only is the scavenging reaction rate reduced compared with ambient storage conditions, but the permeability of the plastic to oxygen is reduced with decreasing temperature. These changes in rates, however, parallel the probable corresponding decreases in growth rate of microorganisms and of food degradation reactions.

RECENT PATENTS AND PATENT APPLICATIONS

Patent applications for oxygen-scavenging plastics during the late 1990s have followed the trend that was evident in the mid-1990s: non-metal systems replacing metals. Dispersions or solutions of low-molecular-weight substances in polymers may be significant, but reactive polymers are the innovative materials with several contenders.

From 1995 to 1997, Chevron Chemical further consolidated their claim to polymers with oxidizable side chains The side chains are incorporated by copolymerizing acrylates with ethylene, with oxidation catalyzed by transition metal salts. The acrylates can typically contain the benzyl group. In 1995, Chevron applied for simple autooxidation of high-impact polystyrene (HIPS) with a transition metal catalyst. Chevron has announced development of transition metal-catalyzed side chain oxidation.

W. R Grace, now Cryovac Division of Sealed Air Corporation, continued defining their light-activated polyolefin oxidation in 1996 by use of a base with the catalyst. In 1997 and again in 1999, Cryovac addressed the need for zeolites or "functional" barriers to migration of malodorous by-products of the autoxidation reactions (Blinka et al., 1998; Speer et al., 1997).

Suppression of malodorous and other by-products from oxygen-scavenging plastics in general has become a topic for patents, or a major mentionable item in patents, in recent years through 2000. Toppan, Grace, Chevron, and Continental PET have all addressed the subject. Some propose polyamines for acids, some suggest zeolites for organics (especially aldehydes), and others propose second laminated functional plastic barriers to odors but not to oxygen.

Several companies have revisited the oxidizable aromatic nylons with transition metal catalysts. These include Continental PET Technologies in 1998 and 1999, Mitsubishi in 1997, and American National Can Company in 1995. Toppan has applied for patents for metal salt-catalyzed oxidation of polyolefins without any antioxidant present. It seems that years of work by the industry to render plastic packaging odor-free is being swept away by many participants intent on using autooxidation rather than reactions that go to completion rapidly and then stop

In 1996, Australia's CSIRO applied for patents for more sophisticated control of activity of triggered scavenging plastics. Continental Can Company entered the field with polyketone films of the type made by Shell Chemical. These films utilize the oxidative instability coupled with catalysts and ultraviolet irradiation. The use of polyketone films in PET gives some barrier to oxidation products Formed carbon dioxide might be used to good effect, for example, it might be advantageous or beneficial in carbonated-beverage or beer packaging.

ADDITIONAL ACTIVITIES

There appears to be a paucity of activity in supplying materials to independent researchers for objective evaluation and publication. In contrast, there is demand for such materials by researchers. There also appears to be a lack of research sponsored by the manufacturers.

The adhesive labels manufactured by Mitsubishi and Multisorb Technologies may be subject to active promotion to companies originally dissatisfied with the concept of loose sachets, as well as to the sachet user network.

CONCLUSION

Oxygen-scavenging plastics are still in their infancy in the United States and worldwide, except in Japan. The patent literature indicates a widespread commercial interest in their potential. Seemingly simple systems based on the blending of low-molecular-weight reagents can present regulatory problems. The control of oxygen reactivity by use of an activation step, such as by use of ultraviolet light, appears to offer a convenient way of overcoming such a limitation. Control of by-products, and especially of potentially toxic migrants, is currently being seen as a most important challenge.

BIBLIOGRAPHY

Abe, Y. 1991. "Oxygen Absorbers: It Is the Answer to Shelf Life Problems?" *Asia Food Industry*, 3(5):66–69, May.

Anonymous. 1988. "Oxygen Absorbing Long Life Military Technology Could Lead to Fresher Longer-Lasting Foods and Beverages." *Good Packaging*, 49(4):22.

Anonymous 1989 "From Japan . . . Yet Another Oxygen Absorber-Preservative." *Packaging Strategies*, August 15.

Anonymous. 1993a. "Scavenging for Safety." *Packaging Week*, 8(42):31

Anonymous. 1993b. "Oxygen Absorbing Sachets Increase Shelf Life" *Food Safety & Security*, 7–8. June

Anonymous. 1994a. "Tiny Pouches Extend Nutraceuticals Lifespan." *Packaging Digest*, 31:8 August.

Anonymous. 1994b. "Additional Oxygen Scavenging Patent (Salicylic Acid Chelates as Oxygen Scavengers Allowed to Advanced Oxygen Technologies)." *TAPPI Journal*, 77:9. September

Anonymous. 1996 "Oxygen-Absorbing Packaging Materials Near Market Debuts." *Packaging Strategies—Supplement.* Packaging Strategies, Inc., West Chester, Pennsylvania. January 31.

Anonymous. 2000. "Hormel Switches to Spin-Welded Plastic Lid for Retorted, MW Bowls for Kids' Meals " *Packaging Management Update*, April 10.

Aoki, T., Y. Himeshima, S Isamu, A. Shirakura, T. Uemura, and Y. S. Yamada. 1991. "Oxygen scavenger; oxygen absorbent component and a permeable film covering the composition " U.S. Patent 5,143,763. September 2.

Bansleban, Donald A., Michael L. Becraft, Thomas A. Blinka, Nathanael R. Miranda, and Drew V. Speer. W. R. Grace & Co.—Conn. 1997. "Functional barrier in oxygen scavenging film." International Patent Application. September 12.

Belcher, J. 1993. "Optimizing the Potential of Oxygen Absorber Technology." *Pack Alimentaire '93.*

Blinka, Thomas Andrew, Drew Ve Speer, and William Alfred Feehley, Jr. 1998. "Oxygen scavenging metal-loaded ion-exchange compositions." U S. Patent 5,798,055. August 25.

Blinka, Thomas A., Christopher Bull, Charles R. Barmore, and Drew V. Speer. 1996. "Method of detecting the permeability of an object to oxygen." U.S. Patent 5,583,047. December 10.

Blinka, Thomas A., Frank B. Edwards, Nathanael R. Miranda, Drew V Speer, and Jeffrey A. Thomas. 1998. "Zeolite in packaging film." U.S. Patent 5,834,079. November 10

Brody, A. L. and E. R. Strupinsky. 1994. "Reduction of Oxygen within Internal Package Environments through Active Packaging and Especially Oxygen Scavengers." Rubbright●Brody, Inc., Report to U.S. Army Natick Research, Development and Engineering Center September 20.

Cahill, Paul J. and Stephen Y Chen. 1998. "Oxygen scavenging condensation copolymers for bottles and packaging articles." International Patent Application WO98/12244. Amoco Corporation. March 26.

Cahill, Paul James, Donald F. Ackerley, Roman F Barski, Jr., Weilong Chiang, David C. Johnson, Walter M Nyderek, George Edmund Rotter, and Stephen Y Chen. 1998. "Zero oxygen permeation plastic bottle for beer and other applications." International Patent Application: WO 98/12127 A 19980326.

Ching, Ta Yen, Joseph L. Goodrich, and Kiyoshi Katsumoto. 1997. Chevron Chemical Company. "Oxygen scavenging system including a by-product neutralizing material." International Patent Application PCT/US97/03307. September 12.

Ching, Ta Yen, Kiyoshi Katsumoto, Steven P. Current, and Leslie P. Theard. 1999. "Compositions having ethylenic backbone and benzylic, allylic, or ether-containing side-chains, oxygen scavenging compositions containing same, and process for making these compositions by esterification or transesterification of a polymer melt." U.S Patent 5,859,145. January 12.

Ching, Ta Yen, Kiyoshi Katsumoto, Joseph L. Goodrich, and J. Diores Gallet. 1996. Chevron Chemical Company. "Oxygen scavenging structures having organic oxygen scavenging material and having a polymeric selective barrier" International Patent Application WO 96/08371. March 21.

Chuu, Michael S. and T. Tung. 1997. "Oxygen scavenging composition." U.S. Patent 5,605,996 February 25.

Cilento, Rudolfo and John Hill. 1978. "Method of using drying oils as oxygen scavenger." U S. Patent 4,073,861. February 14

Collette, Wayne N. and Steven L Schmidt. 1998. "Oxygen scavenging composition for multilayer preform and container." U S. Patent 5,759,653 June 2.

Collette, Wayne N , Steven L. Schmidt, and Suppayan M. Krishmakumar 1998. "Method of forming multilayer preform and container with low crystallizing interior layer." U.S Patent 5,759,656. June 2.

Copeland, J. C., H I. Adler, and W. D. Crow 1991 "Method and composition for removing oxygen from solutions containing alcohols and/or acids." U S. Patent 4,996,073 February 6.

Cullen, John and Samuel Incorvia. 1991 "Packet for compound treatment of gases." U S Patent 5,069,694. December 3.

Farrell, Christopher and Boh C. Tsai. 1985. "Oxygen scavenger" U S Patent 4,536,409 August 20

Frandsen, Erik and Rolando Mazzone. 1993. "Polymer composition for scavenging oxygen." U S Patent 5,194,478 March 16.

Fujinam, Kau, Y. Inoue, S. Murabayashi, A. Natio, and I. Yoshino 1994. "Oxygen absorbent composition and method of preserving article with same" U.S. Patent 5,286,407 February 15.

Gabrielle, Michael C. 1999 "Oxygen Scavengers Assume Greater Role in Food Packaging." *Modern Plastics*, 76(10):73.

Graf, Ernst. 1994 "Copper (II) Ascorbate: A Novel Food Preservation System" *Journal of Agriculture and Food Chemistry*, 42:1,616–1,619

Harimi, Y. 1989 "Ageless Oxygen Absorber." *JPIJ*, 27(6):4–13.

Ho, Y. C., K. L Yam, S. S Young, and P F Zambetti. 1994. "Comparison of Vitamin E, Irganox 1010 and BHT as Antioxidants on Release of Off-Flavor from HDPE Bottles." *Journal of Plastic Film Sheeting*, 10(3):194–212. July.

Hoojjat, P., B. Honte, R. Hernandez, J. Giacin, J., and Miltz. 1987. "Mass Transfer of BHT from HDPE Film and Its Influence on Product Stability" *Journal of Packaging Technology*, 1(3):78.

Hopkins, Thomas R , V. J. Smith, and D. Banasiak. 1991 "Process utilizing alcohol oxidase." U.S Patent 5,071,660. December 10.

Huige, N. 1999. "The Detrimental Effect of Oxygen on Beer Flavor Stability" *Proceedings of "Oxygen Absorbers: 2000 and Beyond" Conference*, Chicago, Illinois, George O. Schroeder Associates, Inc., Appleton, Wisconsin. June.

Ishikzaki, Y. 1999. "The Oxygen Scavenging Technology 'Oxyguard™.'" *Proceedings of "Oxygen Absorbers: 2000 and Beyond" Conference*, Chicago, Illinois, George O. Schroeder Associates, Inc., Appleton, Wisconsin. June.

Katsumoto, Kiyoshi and Ta Yen Ching. 1998. "Multi-component oxygen scavenger system useful in film packaging." U.S. Patent 5,776,361 July 7.

Katsumoto, Kiyoshi, Ta Yen Ching, Joseph L. Goodrich, and Drew Ve Speer. 1998. "Photoinitiators and oxygen scavenging compositions." International Patent Application WO 98/51758. November 19.

Katsumoto, Kiyoshi, Ta Yen Ching, Leslie P. Theard, and Steven P Current. 1996. "Multi-component oxygen scavenger system useful in film packaging." International Patent Application WO 96/25058. August 22

Kim, Yong J and Ray Germonprez. 1994. "Barrier compositions and film made therefrom having improved optical and oxygen barrier properties." U.S. Patent 5,314, 987 May 24.

Klein, T. and D Knorr. 1990. "Oxygen Absorption Properties of Powdered Iron " *Journal of Food Science,* 55(3):869–870.

Kline, L. 1999. "Oxygen Absorber Technology for U S. Military Applications." *Proceedings of "Oxygen Absorbers: 2000 and Beyond" Conference,* Chicago, Illinois, George O. Schroeder Associates, Inc., Appleton, Wisconsin. June.

Komatsu, Toshio and Yokio Kondoh. 1984. "Oxygen absorbent packaging" U S Patent 4,487,791 December 11.

Kondoh, Y 1991. "Japanese perspective on oxygen scavengers." *Pack Alimentaire '91.*

Koyama, Masayasu, Y. Oda, and M. Yamada. 1993. "Oxygen-absorbing resin composition containing water-absorbing polymer, olefin resin and oxygen scavenger." U.S Patent 5,274,024. December 28.

Labuza, T. P. 1990. "Active Food Packaging Technologies." *Food Science & Technology Today,* 4(1):53–56.

Laermer, S F, S. S Young, and P. F Zambetti. 1993. "Vitamin E—New Applications in Packaging Technology." *Proceedings, Future Pack '93*

Leaversuch, R D. 1992 "O$_2$ Scavengers May Replace Barrier Resins in Core Layers." *Modern Plastics,* 69(4):39.

Massouda, D. F 1992. "Oxygen and flavor barrier laminate for liquid packaging." U.S Patent 5,116,649, May 26.

Miller, A. R. and D. A. L. Seiler. 1982. "Oxygen absorbent sachets for preserving bakery products " *British Flour Mill Baking Industry Research Association,* 1(82): 35–45.

Morgan, Charles R., W. P. Roberts, and D. V. Speer. 1993. "Methods and compositions for oxygen scavenging; food and beverage packaging." U.S Patent 5,211,875.

Moschitto, P. 1999. "'TRI-SO$_2$RB in Effect' How Does It Modify your Headspace." *Proceedings of "Oxygen Absorbers: 2000 and Beyond" Conference,* Chicago, Illinois, George O Schroeder Associates, Inc., Appleton, Wisconsin. July.

Nakao, Kiyohiko, Toshio Kawakita, Takanori Kume, and Masashi Sugiyama. 1992. "Oxygen absorbing sheet." U.S. Patent 5,089,323. February 18.

Nawata, Takanari, T. Komatsu, and Y. Kondoh. 1982 "Oxygen absorbent-containing bag." U.S. Patent 4,332,845. June 1.

Newcombe, K. 1991. "Oxygen Absorbers Coming of Age?" *Conference on Shelf Life Problems, Technology and Solutions.* Campden Food & Drink Association, Chipping Campden, U K. March.

O'Keefe, M. and D. Hood. 1981. "Anoxic Storage of Fresh Beef." *Meat Science,* 5:1, 27.

Otsuka, Sadao, Takehiko Moriya, Toshio Komatsu, and Akira Katada. 1986. "Oxygen absorbent packet." U.S. Patent 4,579,223. April 1.

Powers, T. 1999. "'Designing In' an oxygen absorbers," Proceedings of "Oxygen Absorbers: 2000 and Beyond," Conference, Chicago, Illinois, June.

Randell, K., E. Hurme, R. Ahvenainen, and K Latva-Kala. 1995. "Effect of Oxygen Absorption and Package Leaking on the Quality of Sliced Ham." In: P. Ackermann,

M Jägerstad, and T. Ohlsson (eds.). *Food and Packaging Materials—Chemical Interactions.* Royal Society of Chemistry, Cambridge, UK. 211–216.

Reynolds, Patrick. 2000. "Plastic's Beckoning to Beer" *Packaging World*, 7(4):48, 84.

Rooney, Michael L (ed.). 1995. *Active Food Packaging.* Glasgow, UK, Blackie Academic & Professional.

Ryall, G. 1990 "Superbarrier—A Total Oxygen Barrier System for PET Packaging." *Barrier Pack '90*, The Packaging Group, Miltown, New Jersey

Sakakibara, Yoshihisa. 2000. "Evolution of Ageless® Oxygen Absorber—Past, Present and Future" *Proceedings, International Conference on Active and Intelligent Packaging*, Campden & Chorleywood Food Research Association, U.K September.

Sakamaki, C., J I Gray, and B. R. Harte. 1988. "The Influence of Selected Barriers and Oxygen Absorbers on the Stability of Oat Cereal during Storage." *Journal of Packaging Technology*, 2(3) 98–102. June.

Sayer, G. O. 1991. "Package Oxygen Absorbers." *Food Australia*, 43.11.

Schmidt, S. L. and W. N Collette. 1996. "Transparent package with aliphatic polyketone oxygen scavenger." PCT Application US/95/16103. International Publication Number WO 96/18686. June 20.

Schmidt, Steven L., Amit S. Agrawal, and Ernest A Coleman. 1997. "Transparent oxygen-scavenging article including biaxially-oriented polyester." International Patent Application: PCT/US97/16825.

Schmidt, Steven L., Suppayan M. Krishnakumar, and Wayne N. Collette. 1998 "Multilayer container resistant to elevated temperatures and pressure, and method of making the same." U.S. Patent 5,804,016 September 8.

Schroeder, G. M. 1999a. "An Introduction to Oxygen Scavengers: Market Needs, Benefits, Classifications, and Chemistries." *Proceedings of "Oxygen Absorbers: 2000 and Beyond" Conference*, Chicago, Illinois, George O. Schroeder Associates, Inc , Appleton, Wisconsin. June.

Schroeder, G. M. 1999b "Licensable Oxygen Scavenger Technologies from Japan" *Proceedings of "Oxygen Absorbers: 2000 and Beyond" Conference*, Chicago, Illinois, George O. Schroeder Associates, Inc., Appleton, Wisconsin. June.

Shorter, A J. 1982. "Evaluation of Rapid Methods for Scavenging Headspace Oxygen in Flexible Pouches." *Lebersmittel-Wissenschaft und Technologie*, 15(6):380–381.

Spaulding, M. 1988. "Oxygen Absorbers Keep Food Fresher." *Packaging* (U.S), 33(1):9–10 January.

Speer, Drew Ve and William P. Roberts. 1994. "Oxygen scavenging compositions for low temperature use." U.S. Patent 5,310,497, May 10.

Speer, Drew V., William P. Roberts, Charles R. Morgan, and Cynthia L. Ebner 1994a. "Compositions, articles and methods for scavenging oxygen" U.S Patent 5,346,644. September 13.

Speer, Drew V., Charles R. Morgan, William P. Roberts, and Andrew W. VanPutte. 1994b. "Multilayer structure for a package for scavenging oxygen." U.S. Patent 5,350,622. September 27.

Speer, Drew V., William P. Roberts, Charles R. Morgan, and Cynthia L. Ebner. 1997. "Packaging articles suitable for scavenging oxygen." U S. Patent 5,700,554. December 23.

Suzuki, H. et al. 1985. "Effects of Oxygen Absorber and Temperature on Omega 3 Polyunsaturated Fatty Acids of Sardine Oil during Storage." *Journal of Food Science*, 50(2):358–360.

Teumac, F. N. 1994. Personal communication. May.

Teumac, Fred N. 1995. "The History of Oxygen Scavenger Bottle Closures." In: Michael L. Rooney (ed.). *Active Packaging*. Glasgow, UK, Blackie Academic & Professional.

Thomas, J. A. and R. K. Espinel, 1998 "A Polymeric Oxygen Scavenging System and Its Packaging Application." Poster Session 77D-6, *Institute of Food Technologists Annual Meeting*, Atlanta, Georgia.

Tokuoka, K , T. Hirata, T Ishikawa, T. Ishitani, and K. Shinohara. 1991. "Growth of Yeasts in the Packaging Containing Oxygen Absorber." *Japan Packaging*, 12(1):1–13.

Toppan Printing Company. 1992. "Oxygen absorbent compositions which function in dry atmospheres." Japan Patent Application 4298231.

Toyo Seikan Kaisha Ltd. 1995. "Laminate containing an oxygen scavenger" Japan Patent Application 7-67594 March 14.

Trottier, Robert, L. 2000 Personal communication.

Tsai, Boh C. 1996. "Amosorb®: Oxygen Scavenging Concentrates for Package Structures." *Proceedings of Future Pack '96*

Tsai, Boh, A. C. Nielsen, and J. Palm. 1999. "Effect of CO_2 on the Performance of Amosorb® 2000." *Proceedings of "Oxygen Absorbers: 2000 and Beyond" Conference*, Chicago, Illinois, George O. Schroeder Associates, Inc., Appleton, Wisconsin. June.

Venkateshwaran, Lakshmi N., Dinesh J. Chokshi, Weilong L. Chiang, and Boh Chang Tsai. 1998. "Oxygen-scavenging compositions and articles." U.S. Patent 5,744,056. April 28.

Watanabe, M. 1999. "Looking Back and Looking Forward . . . Market Demands in Japan and North America and the Evolving Nature of AGELESS® Oxygen Absorbers." *Proceedings of "Oxygen Absorbers. 2000 and Beyond" Conference*, Chicago, Illinois, George O. Schroeder Associates, Inc., Appleton, Wisconsin. July.

White, S. 1999. "Why Plastic Beer Bottles Need Oxygen Absorbers." *Proceedings of "Oxygen Absorbers: 2000 and Beyond" Conference*, Chicago, Illinois, George O. Schroeder Associates, Inc., Appleton, Wisconsin. June.

Wright, R. V. and A M. Chuprevich. 1994. "Method for extending shelf life of juice" U.S. Patent 5,324,528.

Zenner, B., J. P. Ciccone, E. S. De Castro, L. A. Deardurff, and J B Kerr 1990. "Polyalkylamine complexes for ligand extraction and generation." U.S. Patent 4,959,135 September 25.

CHAPTER 5

Moisture Control

W ATER loss from fresh produce, fresh meat, fish, poultry, or other fresh or minimally processed prepared foods as a result of normal respiration, microbiological activity, or physical activity can occur as a result of evaporation from the product followed by permeation through the package material when the package material does not provide an adequate water-vapor barrier The use of polyethylene film carton and case liners for fresh produce, meat, poultry, fish, and prepared food often satisfies the short-term commercial moisture-barrier requirement but may not provide sufficient packaging if temperature cycling occurs during the distribution of the product. When the temperature of the product drops slightly, condensation of water-vapor deposits liquid water on the inside of the liner. If the water-vapor content of the package headspace is buffered to a relative humidity substantially below the approximately 95% level generated by respiration of fresh product, a safety margin exists before condensation occurs. This feature is a characteristic of recent research on paperboard carton coating in which the product is kept away from a hydrophilic polymer carton liner by means of a porous polymer. Water vapor can pass from the headspace atmosphere into the hydrophilic polymer as the temperature varies during distribution.

Condensation or "sweating" is a problem in many kinds of packaged foods, particularly fresh fruit and vegetables. When one part of the package becomes cooler than another, water vapor condenses as liquid droplets in the cooler areas. If the liquid water is kept away from the product, the harm is package appearance and consumer appeal, both of which are important. However, when condensation moistens the product's surface, soluble nutrients leak into the water, encouraging rapid growth of ubiquitous mold spores and leading to loss of nutrients from the product.

Several desiccants such as silicates (i.e., silica gel) and humidity-controlling salts have long been used in food packaging, particularly to maintain the relative humidity surrounding dry foods and non-food products such as pharmaceuticals. The amount of silica gel, its moisture-absorbing rate, and the water-vapor transmission rate of a water-vapor-permeable sachet membrane control the moisture content within the package, although usually at a very low relative humidity level.

RELATIVE HUMIDITY CONTROL WITHIN PACKAGES

The use of humidity-control technology reduces condensation inside packages of respiring and other high-water-content foods and eliminates water films on the food without further drying the food. Therefore, moisture-sensitive products such as fresh-cut lettuce, fruit, etc., are protected from contact with water. This action helps reduce surface mold growth and thus extends microbiological shelf life. Some moisture-control active packaging systems are described here.

DOW BRANDS

During the early 1990s, through its Dow Brands operations, Dow Chemical introduced Summerfield-brand, fresh whole tomatoes in hermetically sealed thermoformed plastic trays. Dow Chemical developed the packaging technology and then marketed the tomatoes using the technology. The internal relative humidity of the tray was controlled by the presence of sodium chloride. The salt was ultrasonically heat-sealed within high-water-vapor-permeability Tyvek® spun-bonded polyolefin film sachets. The salt was intended to control internal relative humidity to levels of 80% to 85% to retard significant moisture loss from the tomatoes and, further, to retard surface mold growth. The system functioned well technically but was removed from the marketplace for marketing reasons.

CHEFKIN

Japan's Chefkin has developed package overwrap material claimed to be capable of controlling the relative humidity within a package. This overwrap consists of a duplex of two sheets: the external sheet is a water-vapor barrier and the inner sheet is a water-vapor-permeable (but not water-permeable) film. Together they create a pocket. Between the films is an interior void containing a liquid glucose solution. When the interior of the package has a high relative humidity, water passes from the package contents through the inner film into the glucose solution. When the contents exhibit low relative humidity, water passes in vapor form from the glucose solution into the package interior. The glucose concentration of the trapped solution determines the precise relative humidity at which the water-vapor passage occurs. The principle would imply that almost any reasonable relative humidity may be controlled within a package through the use of a liquid trapped in a high-water-vapor-permeability package overwrap structure, with a water-vapor barrier on the exterior.

The company claims that the material is useful for wrapping fish and "fresh dry" products and that the overwrap material can be used repeatedly, implying that the liquid's solute concentration may be reversed (Matsui, 1989).

HUMIDOR BAG

The flexible pouch called the Humidor Bag has been manufactured in the United States since 1998, and is intended to maintain the 70% relative humidity required for quality retention of cigars. According to the patent and related information, the pouch, which has a zipper closure, uses System 70™ Humidi-Pak™ technology. An external pouch contains a "natural solute solution" that maintains a precise 70% relative humidity within the four-side-sealed pouch to which it is attached. One face of the pouch is an impermeable aluminum foil lamination, and the other face is a transparent water-vapor-permeable (but not liquid-water-permeable) film, which also functions as one face of the pouch containing the "solute solution." The consumer opens the pouch and places the cigars in the inner chamber that has one moisture-impermeable aluminum foil lamination face and one water-vapor-permeable face, containing the liquid. The pouch zipper is then closed to develop the desired relative humidity. Obviously, this technology could be applicable for food packaging.

PITCHIT

Japan's Showa Denko Company developed Pitchit Film to decrease water activity at the food surface. Propylene glycol produces the "active packaging" action. The package is a sandwich composed of two sheets of polyvinyl alcohol (PVA) film sealed along the edge. Between the two sheets is a layer of propylene glycol humidifying agent. The PVA film is very permeable to water-vapor but is a barrier to the propylene glycol.

The laminated product is sold directly to food consumers with the PVA blanket and 10 sheets of cellophane film. The consumer is instructed to put a sheet of cellophane around the food to be preserved (whole fish, fish fillets, other seafood, meat, poultry) and then wrap the blanket around that. The unit is then kept in the refrigerator for 4 to 6 hours during which time the surface of the food is dehydrated. The action is due to a water-activity difference between the food (a_w 0 99) and the propylene glycol; thus water is rapidly drawn from the food surface. This surface dehydration not only inhibits some microorganisms but also may injure others without causing change in product quality. It is possible that some propylene glycol transfers to the food surface, also retarding microbial growth. The package reportedly provides an extra 3 to 4 days of refrigerated shelf life to fish or other contents. After removal, the company claims the blanket can be washed in water, dried, and reused up to ten times (Miyake, 1991).

DAI NIPPON

Japan's Dai Nippon offers a water-adsorptive polymer such as sodium polyacrylate, plus an antimicrobial and/or deodorant material such as charcoal powder, and a thermoplastic binder such as ethylene vinyl acetate spread on a fluid-adsorptive sheet such as paper, and covered with another sheet of paper. The combined adsorptive assembly is sandwiched between a water-impermeable sheet, such as polyethylene-laminated paper, and a nonwoven fabric, such as a polyolefin. The multilayer sheet is embossed with a hot roll to fuse the layers together. The sheet is used as a food-wrapping material, with moisture passing into and being trapped by the fluid-adsorptive sheet. This idea is not far removed from absorptive pulp pads used as soaker pads to remove liquid purge from the vicinity of the product in retail meat trays (Mita et al., 1990).

KYUSHI

A drip tray has been jointly developed by Japan's Dai-ichi Plastic Industry Company and Dai Nippon. The Kyusui Tray is fabricated from polypropylene, and at the bottom of the tray is a water-adsorbing sheet covered with a nonwoven fabric closed with pressure-sensitive adhesive. This construction prevents the drip fluid from leaking when the customer carries the filled tray home in a carrier bag, even if the tray is at an angle or upside down. It also saves supermarkets the labor of putting soaker pads into each tray when packaging meat, fish fillets, and other foods that tend to squeeze out liquid (Anonymous, 1992).

CRISPER

"Crisper F" is a sheet product for food packaging, made of aluminum metallized film with nonwoven fabric on the reverse side, to absorb meat and fish exudations. Developed by Japan's Kagaku Kogyo, the Crisper series of materials is also claimed to be suitable for packaging fresh fruit and vegetables. It retards deterioration by absorbing liquid and controlling relative humidity. Water evaporation is slowed and discoloration retarded when applied to fruit packaging. For this application, film pouches and a tight seal are necessary. Carbon dioxide and ethylene gas are also claimed to be absorbed.

GRACE CONDENSATIONGARD®

Among the United States' attempts to control the relative humidity within closed packages is Grace Davison's CondensationGard®. According to the developer, a conventional desiccant such as silica gel continuously adsorbs water-vapor at low levels of relative humidity until the desiccant is saturated and, possibly, the contained product is partially dehydrated. Condensation-Gard® does not desiccate a product that requires a moist environment. This adsorbent product does not become active until the package's environment reaches the condensation point, at which time CondensationGard® adsorbs only excess moisture, eliminating condensation while retaining the proper moisture balance in the product.

In a sealed packaging environment, moisture contained in the atmosphere and in the food product is trapped within the package. As the temperature falls during shipment or refrigeration, the relative humidity rises. When the relative humidity reaches the dew point (100% relative humidity), condensation occurs.

CondensationGard® silica gel differs from conventional desiccants in that it has very little capacity to adsorb moisture below 50% relative humidity (about 7% by weight) and will adsorb only about 11% of its own weight at 80% relative humidity. At the dew point, however, the material has a total moisture capacity of over 90%. This means that CondensationGard® adsorbs more than 80% of its own weight as the relative humidity rises from 80% to 100%. This unique property imparts to the material the ability to adsorb water-vapor just below its dew point, thus removing moisture from the air before condensation can occur.

The active component is packaged in a highly gas-permeable, nonwoven polyolefin bag and can be produced in a range of sizes. CondensationGard® is available in two grades, which respond to different relative humidity levels, 90% and 75% relative humidity.

OTHER TECHNOLOGIES

Materials such as polyacrylates surrender adsorbed water into both plastic and plastic packaging contents. These materials are employed as active water adsorbers in retail fresh meat tray soaker pads in Europe, but are not accepted by the FDA in the United States for what the FDA regards as direct food contact.

A recent alternative involves the use of a paperboard carton as the active package. An integral water-vapor barrier layer is on the inner surface of the fiberboard. A paper-like material bonded to the barrier acts as a wick, and a layer highly permeable to water-vapor, but not to water, is next to the contained fresh fruit or vegetable. The latter water-vapor-permeable layer is spot-welded to the layer underneath. The multilayer material on the inside of the carton is able to take up water in the vapor state when the temperature drops and the relative humidity rises. When the temperature rises, the multilayer material releases water-vapor back into the carton in response to a lowering of the relative humidity. The condensation-control system therefore acts as an internal water buffer. The critical characteristic of the system is the capacity of the wicking layer for water.

MOISTURE REMOVERS

Desiccants protect sensitive products against water and humidity. Offered in clay, molecular sieve, and silica gel forms, they absorb moisture that enters or remains in a package.

Probably the best known internal package moisture absorbers are conventional silica gels, which can absorb up to 35% of their own weight in water. Silica gels are useful to maintain dry conditions within packages of dry foods, down to below 0.2 water activity.

Molecular sieves such as zeolites can absorb up to 24% of their weight in water, i.e., they have a high affinity for moisture. When dry, they can also pick up odors. When wet, they tend to surrender these odors.

Cellulose fiber pads are often used to remove water, as, for example, soaker pads in the bottom of meat, poultry, and fresh produce trays. They can also surrender water when they are saturated. As such, cellulose fiber pads can be employed to add water to the internal package environment.

MOLDING DESICCANTS INTO PACKAGE MATERIALS

Traditional technologies impose limits on desiccant use and cost. The current packaged forms occupy too much interior space within packages. Packaged desiccants restrict the ability of package structural designers to place the material at optimal locations within a package. Moreover, secondary (often manual) insertion of packaged desiccants slow packaging lines and add cost. One answer to these concerns is to compound, mold, or extrude the desiccant into the plastic melt much like a filler or additive is currently incorporated. In such technologies, the desiccant is blended in a polymer melt, reportedly without degrading the performance of either the plastic or the desiccant. Technology is also needed to make the polymer carrier "breathable," so that the desiccant in the plastic is still able to absorb moisture. Finally, a way to prevent moisture from penetrating from the outside (e.g , using an aluminum foil or other barrier) is necessary.

Technologies have been commercialized for molding a desiccant directly into plastic packaging to better protect moisture-sensitive contents. These technologies allow a plastic package converter to "entrain" a desiccant in the polymer melt stream for molding into a container wall or into the bodies of closures or sleeve inserts. This approach is reported to work better than packaged desiccant methods in which the desiccant is encapsulated in a sachet, tablet, canister, or tube, and then inserted into the package in a secondary step.

United Desiccants, a Süd-Chemie Performance Packaging subsidiary, is one developer of moisture-removing packaging. Süd-Chemie's desiccant technology was developed by France's Airsec, a plastics molder specializing in pharmaceutical applications. Airsec reported an initial commercial application in Europe for a medical diagnostic kit. They introduced desiccant-filled cap liners that could be inserted into standard screw-on bottle closures. Container Dri® desiccant was developed specifically to combat condensation during transport.

Another developer of molded desiccants is Capitol Specialty Plastics, Inc., an injection molder of pharmaceutical and food containers. Capitol can integrate desiccants into product packaging such as vials, canisters, containers, film, string, sheets, trays, or strips. Specific adsorption desiccant-delivery systems can have varied adsorption rates and capacities. Capitol uses a variety of desiccants, silica gel and others, and their materials are FDA accepted. Desiccant film is produced in a variety of widths and thicknesses; the film changes color from blue to pink as adsorption capacity is reached. Capitol desiccant can be created from virtually any plastic material in any shape and size by forming processes including injection molding, co-injection molding, thermal forming, and blow molding. Custom injection molding is used to create vials, canisters, containers, tablets, wafers, strands, film, sheets, trays, strips, pills, and beads.

Market potential for the new approach in desiccants is reported to exist in packaging for medical diagnostic and test-strip kits, effervescent drugs, and nutritional supplement products. At least one test kit incorporating a silica gel-containing plastic is on the market in Europe, using materials made in the United States. The various groups reportedly plan to apply the new approach to moisture-control design by collaborating with their user customers to marry desiccants with the plastics-molding process Obviously, if such products can be employed for packaging drugs, they may also be applicable to foods (Leaversuch, 1997).

CONCLUSION

In-package moisture control is usually targeted at either very dry foods or at respiring, wet, and consequently high relative humidity, fresh/minimally processed produce. To date, silica gel has been the moisture absorbent of choice for dry foods, but incorporated into sachets or other forms placed within the packages. More recently, several cooperative developments among desiccant and plastics processors have resulted in incorporation of desiccant into the plastic being extruded and/or molded. These emerging technologies have potential application to maintain dry conditions within packages of very low water activity foods, i.e., dry foods with water activities of 0.7 and below.

BIBLIOGRAPHY

Anonymous 1990. "Self-Sealing Drip 'Tray'." *Hoso Times,* No. 1511. September 3.

Anonymous. 1992. "Drip 'Absorbent' Sheet and Freshness Retention Agent." *Packaging Trends Japan,* 92:5. November 11.

Leaversuch, Robert. 1997. "Desiccant Molding Technique Builds Moisture Resistance." *Modern Plastics,* May.

Mita, K., J. Kuroki, and J. Inaoka. 1990. "Water-adsorptive sheet." Japanese Patent Kokai 252,554/90. October 11.

Matsui, M. 1989 "Film for keeping freshness of vegetables and fruit." U.S. Patent 4,847,145. July 11.

Miyake, T. 1991. "Food wrap sheet." Japanese Patent Kokai 65,333/91. March 20

Patterson, B. D., J. J. Jobling, and S. Moradi. 1993 "Water Relations After Harvest— New Technology Helps Translate Theory into Practice " *Proceedings of Australian Postharvest Conference,* Gatton, Australia, September 20–24.

Wagner, B. and N. Vaylen. 1990. "The Packaging Activists." *Prepared Foods,* 159(9): 172.

CHAPTER 6

Gas-Permeability Control

Not yet generally listed among the major categories of active packaging are plastic materials that are sensitive to temperature and that significantly alter their gas permeabilities as a consequence

AVOIDANCE OF ANOXIC CONDITIONS

In modified atmosphere packaging of fresh and minimally processed foods, oxygen is often intentionally reduced to decrease enzymatic, biochemical, and aerobic microbiological activities. One problem with reduced oxygen is that if the oxygen approaches or reaches the zero level, undesirable reactions may occur. In the case of respiring fruit and vegetables, anaerobic respiration occurs along pathways different from aerobic respiration and produces odoriferous aldehydes and ketones and other malodorous compounds. With low-acid foods, under zero-oxygen conditions, pathogenic anaerobic microorganisms may grow and produce toxins. Thus, although low oxygen is desirable to prolong quality, oxygen extinction is often undesirable and may even be hazardous.

To obviate this problem, package material converters have developed and are continuing to develop high-gas-(i.e., oxygen)permeable package materials. None of the "conventional" commercial high-gas-permeability materials, however, is capable of changing gas permeability with increasing temperatures. During commercial distribution, temperatures can rise from the refrigerated to well above optimum, with consequential increases in the activity, i.e., oxygen consumption, levels of the contained food but without associated increases in gas permeabilities of the package materials. To overcome this anomaly, several organizations have been developing package materials and adjuncts that are sensitive to temperature, i.e., that significantly alter their gas permeabilities with increasing temperatures.

TEMPERATURE-SENSITIVE STRUCTURES

DIFFERENTIAL COEFFICIENT OF EXPANSION

In the United Kingdom, a two-ply film containing small cross-cuts across the surface was developed, but not commercialized. Each ply had a different coefficient of expansion with increasing temperature, and so, as temperature increased, the points tended to curl in the direction of the more rapid expansion, increasing the opening at the cross-cut. Thus, the volume of oxygen-containing air passing in from the exterior was increased. When the temperature decreased back to the desirable refrigerated level, the "bimetallic" effect relaxed, and the film returned to its flat condition, closing the orifices to further gas transmission. Thus, in theory, this technology was reversibly temperature sensitive.

MELTING WAX

Among the temperature-sensitive offerings and developments are and/or have been films containing pores filled with waxes, which melt and flow out of the pores when the temperature increases. This flow leaves open pores through which atmospheric oxygen can flow. This response is irreversible.

SIDE-CHAIN CRYSTALLIZABLE POLYMERS

Probably the most widely known of the temperature-responsive package materials are Landec's Intellipac™ technologies. Landec's technology can provide correct atmosphere and maintain it with temperature through a highly permeable membrane in the wall of the package. The membrane is made by coating a porous substrate with a proprietary Landec side-chain crystallizable (SCC) polymer. Because Landec can modify the properties of the polymer and the membrane, they can achieve the oxygen permeability desired, the correct carbon-dioxide-to-oxygen-permeability ratio, and the required change in permeability with temperature. Conventional films based on polyethylene polymers maintain a relatively fixed carbon-dioxide-to-oxygen ratio. Also, with conventional films, the permeability can ultimately be increased only by reducing the gauge of the film, which consequently reduces the strength of the package structure and the weight of product that can safely be packaged. When using the Intellipac™ membrane, the permeability of a package can be increased by increasing the size of the membrane. This independent control of permeability allows the packaging of very large quantities of produce while still exhibiting the correct permeability and gas atmosphere.

SCC polymers are acrylic polymers with side chains capable of crystallizing independently from the main chain. By varying the chain length of the side chain, the melting point of the polymer can be altered. By making the appropriate copolymers, it is possible to produce any melting point from 0° to 68°C. The SCC polymers are unique due to their sharp melting transition and the ease with which it is possible to produce melting points in a specific temperature range. When heated to the "switch" temperature, SCC polymers become molten fluids. The SCC polymers are intrinsically highly permeable. The polymer properties also can be modified by the inclusion of other monomers to change the relative permeability of carbon dioxide to oxygen, for example.

The Intellipac membrane is made by coating a porous substrate with an SCC polymer to provide a membrane having specific properties. By altering the gauge of the coating, it is possible to alter the permeability of the membrane, allowing the packaging of high-respiration-rate fresh produce without compromising the mechanical integrity of the package. In addition, the package material can be selected to provide the correct water-vapor transmission characteristics.

By altering the polymer composition used to coat the membrane, Landec can produce membranes with carbon-dioxide-to-oxygen-permeation ratios ranging from 18:1 to 1.5:1. A membrane with a carbon-dioxide-to-oxygen ratio of 18:1 results in a package containing relatively little carbon dioxide and is useful for fresh produce, which is sensitive to carbon dioxide. A membrane with a ratio of 1.5:1 results in a package containing higher levels of carbon dioxide A perforated package, however, has a carbon-dioxide-to-oxygen transmission ratio of 1:1, which will produce high levels of carbon dioxide at oxygen levels below 10% oxygen by volume. While elevated levels of carbon dioxide can prevent the growth of many microorganisms, too high a level may cause off-flavors, and so a need exists to be able to adjust the carbon-dioxide-to-oxygen ratio to provide the correct level of carbon dioxide (Bodfish, 1998).

The Landec material is being considered for use with minimally processed prepared foods that are packaged under reduced oxygen for chilled distribution. Under zero-oxygen or anoxic conditions, low-acid prepared foods might be vulnerable to anaerobic microbiological safety issues (Biggs, 1999, 2000).

BIBLIOGRAPHY

Biggs, Lynn. 1999 Personal communication.
Biggs, Lynn 2000. Personal communication.
Bodfish, Lynn 1998 Personal communication

CHAPTER 7

Ethylene Control

ETHYLENE is a growth hormone that functions in the sprouting of plant seedlings, the growth of plants, and the growth of fruit. It helps to accelerate ripening in fruit, followed by aging and ultimately death. Ethylene production is a biochemical process, independent of respiration, that occurs in each living cell for the purpose of producing energy.

The respiration rate of fresh produce varies with type. Respiration rate is characterized by the quantity of carbon dioxide emitted. Apples, kiwis, and grapefruits have low respiration rates, i.e., they emit very little carbon dioxide. Apricots, peaches, pears, plums, bananas, carrots, lettuce, and tomatoes have mid-range respiration rates. Strawberries, raspberries, avocados, cauliflower, and green onions have high respiration rates. Asparagus, broccoli, and mushrooms respire at extremely high rates.

Much fresh produce emits relatively large amounts of ethylene and smaller amounts of alcohol, aldehyde, and ester vapors in addition to carbon dioxide. Ethylene emission differs according to type of plant as well, but high-respiration-rate produce is not necessarily a strong ethylene emitter, and vice versa. For example, apples and kiwis do not respire at high rates but do produce large amounts of ethylene. Potatoes respire slowly and emit small amounts of ethylene, but asparagus respires at a high rate and produces little ethylene. Ethylene production is greatly increased during bacterial or fungal infections of fruits and vegetables, as well as by mechanical and/or chemical damage (stress ethylene).

As ethylene in the atmosphere surrounding produce increases, the plant's respiration rate increases. If the ethylene level of the surrounding environment is maintained low, respiration slows. The effect on the respiration rate depends primarily on the relative concentration compared to the ethylene emission of the plant, rather than on the absolute ethylene concentration.

When ethylene is removed from the fresh fruit or vegetable environment, the ripening and deterioration processes of plant products are slowed, and so the storage life is extended.

POTASSIUM PERMANGANATE REACTION

Ethylene removal media are usually composed of pellets of porous solids such as activated alumina, vermiculite, and silica gel that have been impregnated with potassium permanganate ($KMnO_4$). The alumina functions primarily as the absorptive surface to trap the molecules of gas and is a carrier for the permanganate. Potassium permanganate is a broad-spectrum oxidizing agent that reacts with ethylene as well as with other contaminant gases. When it reacts by oxidizing ethylene to ethylene glycol, a visible color change occurs: fresh $KMnO_4$ medium is purple in color but, after reacting with ethylene, it turns brown. The color change occurs first on the pellet's surface and eventually penetrates the core, indicating the reactive capacity is nearing exhaustion.

Typically, ethylene scrubbing media are packaged in several different forms, including blankets, tubes, and sachets. Sachets are utilized in individual boxes of fruits or vegetables, whereas blankets and tubes are commonly employed in transport vehicles (Lasris, 1990)

Reducing the concentration of the gas surrounding the fruit causes the ethylene that remains inside the fruit to diffuse to the exterior. The result is that less ethylene is available inside the fruit to accelerate maturation.

Ethylene scrubbers are particularly useful when fresh produce is stored or shipped in low-oxygen atmospheres. When applied, modified atmosphere delays ripening. Low-oxygen atmospheres delay ripening due to retardation of ethylene production. However, while in low-oxygen atmospheres, methionine—an ethylene precursor—accumulates in the cells. Methionine is converted to ethylene at an accelerated rate when oxygen is introduced to the system. Without the protection of scrubbers, this reaction series results in a rapid deterioration of the fresh produce.

Carriers for potassium permanganate such as alumina are offered in a variety of grades and sizes, and they differ in pore sizes, pore volumes, surface areas, and bulk densities. In general, the lower the bulk density, the more efficient the resulting product will be. The loftier alumina has larger and a greater number of pores running throughout the individual pellets, resulting in greater surface area available to trap the gas. In use, many pellets do not turn brown all the way through because air cannot penetrate to the inside.

OTHER ETHYLENE SCAVENGERS

A number of alternative ethylene-scavenging systems have been proposed. Activated charcoal alone or after impregnation with bromine is another ethylene-absorbing system. Also, bentonite, Kieselguhr, and crystalline aluminosilicates, e.g., zeolites, have been reported capable of adsorbing ethylene.

In Korea, Orega ethylene scavengers were developed. Film made incorporating the Orega compound is claimed to have an ability to scavenge ethylene at a rate of at least 0.005 ppm per hour per square meter. The ethylene-adsorptive activity of this film results from adding into the film a fine porous, inorganic material containing a large number of fine pores, 2 to 2,800 Å in size, such as pumice, zeolite, active carbon, cristobalite, and clinoptilolite. The finely divided porous material is sintered with a small amount of a metal oxide before being added to the film The fine porous material is then incorporated into the plastic polymer for the film by conventional methods, and the polymer is extruded into the film. The particle size of the fine powder should be at least 200 mesh, and at least 1% by weight should be contained in the film. Treating the porous mineral with oxygen enhances the ethylene-absorbing activity.

The film containing the fine porous material not only has ethylene-scavenging activity, but reportedly also has excellent permeability to gases such as oxygen, carbon dioxide, nitrogen, ethylene, and water vapor. Consequently, ethylene gas is discharged outside the film wrapping vegetables and fruit, and the inside of the film is maintained at a suitable relative humidity (Choi, 1991).

An Austrian company, E-I-A Warenhandels GmbH, produces "ProFresh" ethylene absorber whose active ingredient is not identified. "ProFresh" is a mineral-based ethylene and ethanol absorber. The "ProFresh" product is reported by the supplier to be in commercial use in Italy, Israel, Colombia, Austria, United Kingdom, Hungary, Taiwan, and New Zealand. The application of the masterbatch offered by E-I-A Warenhandels is being developed primarily by flexible package converters. "ProFresh" users have reported good results with tomatoes, paprika, cucumbers, flowers (Israel), cabbage, salads (Hungary), bananas (Colombia), cauliflower, meats, salami (Israel), and apples (New Zealand). In Italy, Austria, and Hungary, "ProFresh"-filled, fresh-keeping, and odor-controlling small bags are sold for retail consumer use to help keep not only the vegetables and fruits fresh, but also to store meat and fish in the refrigerator. Oxidation is reduced and odor is also reportedly reduced. In "ProFresh"-filled trash bags, household rubbish is claimed to not have a bad odor as it usually does after two to three days.

Japan's "Everfresh" film, claimed to have preservative effects, contains the highly publicized "Oya-Stone." The active ingredient of "Everfresh" is not a reproducible compound, but rather a natural stone excavated from a cave in Japan and that includes redundant or inert ballast materials. Oya-Stone is a zeolite with a claimed affinity for ethylene and carbon dioxide. Films containing Oya-Stone have been extruded with polyethylene into pouches used to contain ethylene-generating fruit or vegetables (Anonymous, 1990).

DUNAPACK

A Hungarian paper manufacturer, Dunapack, has developed paper called "Frisspack" for manufacture into corrugated fiberboard cases to prolong the shelf life of contained fresh produce. A component of their "Frisspack" package tries to bond with the ethylene generated during fruit and vegetable ripening in order to decrease the rate of respiration To bond with ethylene, the most commonly used materials are silica gel, with potassium permanganate compounds to react with the ethylene. According to Dunapack, silica gel adsorbs ethylene while $KMnO_4$ oxidizes it.

One of the disadvantages of bag-in-box active packaging systems offered by other companies is that the ethylene-bonding capacity of the materials used as adsorbents (silica gel, activated carbon) is usually low and therefore relatively large amounts of absorbents are required. In addition, adsorbent materials bond with moisture, and the chemicals within the pouch may contaminate the food.

Frisspack paper was developed to chemically bond with ethylene released during fresh and minimally processed produce respiration. Dunapack achieved this objective by dispersing a chemosorbent of small particle size (average particle size 1 μm) with high absorption capacities among the fibers in the early phase of the paper production, when the fibers are still in approximately 1% water suspension. The chemosorbent can be uniformly blended among the fibers and have surface characteristic features such that the particles connect to the paper fibers by hydrogen bonding. By this means, the chemosorbent particles are embedded in the paper so firmly that they do not become separated from the paper due to mechanical action or chemical effects (e.g , juice leakage from fruits during storage).

The absorption capacity is largely dependent on the ethylene concentration of the atmosphere. The higher the ethylene concentration of the atmosphere, the higher the amount of ethylene that can be absorbed by Frisspack paper until equilibrium is achieved. The absorption capacity also depends on temperature: the higher the temperature, the less the amount of absorption. Frisspack paper, however, is capable of absorbing fairly large amounts of ethylene even at high temperature, i e., well above optimal refrigerated temperature.

In a closed environment, the ethylene concentration of the atmosphere rapidly decreases due to the ethylene adsorbed by the paper. The rate of the concentration change does not depend on the original concentration. The chemosorbent does not fundamentally affect the ethylene diffusion through the paper. The ethylene diffusion through Frisspack paper is similar to that of untreated paper. The paper must contain 2 to 4 g/m^2 chemosorbent so that its ethylene-absorbing capacity fits the requirements.

The effect of Frisspack package materials on the ripening of fruits was tested with a wide variety of fresh produce products, including apples, pears, peaches, apricots, bananas, cherries, grapes, strawberries, raspberries, carrots, onions, potatoes, mushrooms, tomatoes, and green peppers. Frisspack paper significantly retarded the ripening of almost all types of fruits and vegetables tested. The loss in weight and sugar content of the fruits stored in packages made of Frisspack paper was less, and they retained their texture and color to a greater extent than the fruits stored in conventional packaging materials (Szikla, 1993).

SACHET ETHYLENE SCAVENGERS

Several ethylene-scavenging systems have been patented and are commercially available. As indicated above, these systems usually involve the inclusion of a small sachet containing the scavenger in the package. The sachet material itself is highly permeable to ethylene, and diffusion is not a limitation. Other systems incorporate the scavenger directly into packaging films. Silica gel and activated carbon can adsorb ethylene gas but do not oxidize ethylene However, they can be used with potassium permanganate to increase the ethylene-removal capacity.

ACTIVATED CARBON ETHYLENE REMOVERS

Metal catalysts on activated carbon remove ethylene from air passing over the bed of activated carbon. Activated charcoal impregnated with a palladium catalyst and placed in paper sachets effectively removes ethylene by oxidation from packages of minimally processed kiwi, banana, broccoli, and spinach (Abe and Watada, 1991).

Japan's Sekisui Jushi has developed a sachet containing activated carbon and a water absorbent that is capable of absorbing up to 500 to 1000 times its weight of water. The company provides data showing that the product, tradenamed Neupalon, absorbs 40 cc ethylene per square meter of package surface area.

Japan's Honshu Paper offers the "Hatofresh Systems," which are based on activated carbon impregnated with bromine-type inorganic chemicals. The carbon-bromine substance is embedded within a paper bag or corrugated fiberboard case used to hold fresh produce. They claim that the bag will scavenge 20 cc of ethylene per gram of adsorbent by the combination of ethylene with bromine.

Japan's Mitsubishi Chemical Company produces "SendoMate," which employs a palladium catalyst on activated carbon that adsorbs ethylene and then catalytically breaks it down. The product comes in woven sachets that can be placed in packages of produce.

ACTIVATED EARTH ETHYLENE REMOVERS

In recent years, a number of packaging products have appeared based on the ability of certain finely dispersed clays and activated earths to adsorb ethylene. Typically, these materials are local clays (often zeolites) incorporated into polyethylene bags, that are then used to package fresh produce. Most of the bags are marketed by Japanese or Korean companies

Japan's Nissho suggests a film that incorporates finely ground coral (primarily calcium carbonate), having pore sizes in the range of 10 to 50 μm. After incorporation into a polyethylene film, the coral is claimed to absorb ethylene.

"BO Film," marketed by Japan's Odja Shoji, is a low-density polyethylene film extruded with finely divided crysburite ceramic, claimed to confer ethylene-adsorbing capacity.

Zagory, in his section on ethylene removal in the book *Active Food Packaging* (Rooney, 1995), asserts that the evidence offered in support of claims about ethylene removal by activated earth or clay is generally based on shelf life experiments comparing common polyethylene film bags with clay-filled film pouches. Such evidence generally shows an extension of shelf life and/or a reduction of headspace ethylene. Although the finely divided clays may adsorb ethylene gas, they can also create pores within the plastic bag and alter the gas-transmission properties of the bag. Because ethylene diffuses more rapidly through open pore spaces within the plastic than through the plastic itself, ethylene would be expected to diffuse out of these pouches faster than through pristine polyethylene film bags. In addition, carbon dioxide within these pouches is transmitted more rapidly and oxygen enters more rapidly than with a comparable conventional polyethylene film pouch due to the gaps in the film. These effects can enhance shelf life and reduce headspace ethylene concentrations independently of any ethylene adsorption. In fact, almost any powdered mineral can confer such effects without relying on expensive Oya Stone or other specialty minerals.

Although the minerals in question may have ethylene-adsorbing capacity, the data supporting the commercial products incorporating these minerals have not demonstrated such capacity. Even if they have ethylene-adsorbing capacity, it is probable that they will lack significant capacity while embedded in plastic films. The ethylene would have to diffuse through the relatively gas-impermeable polymeric matrix before contact with the dispersed mineral, thus greatly slowing any adsorption processes. Once the ethylene has diffused part-way through the plastic film, venting to the outside may be nearly as fast and effective as adsorption by embedded minerals.

NOVEL ETHYLENE-REMOVING PACKAGING

Perhaps the most promising new development in ethylene-removing packaging is the use of electron-deficient nitrogen-containing trienes incorporated in ethylene-permeable packaging. The preferred diene or triene is a tetrazine. However, since tetrazine is unstable in the presence of water, it must be embedded in a hydrophobic, ethylene-permeable plastic film that does not contain hydroxyl groups. Appropriate films would include silicone polycarbonates, polystyrenes, polyethylenes, and polypropylenes. Approximately 0.01 to 1.0 M of the dicarboxyoctyl ester derivative of tetrazine incorporated in such a film was demonstrated to be able to effect a tenfold reduction in ethylene in sealed jars within 24 hours and a 100-fold reduction within 48 hours.

CONCLUSION

In the United States, ethylene control within packages of fresh and minimally processed fruit and vegetable products remains almost exclusively a reaction of potassium permanganate on a porous mineral structure. Despite the many Asian claims made on the effectiveness of activated carbon, natural mineral earths, and other mixtures blended into plastic film, little credible proof has been offered to convince United States package material converters and packagers that these structures are functional. Thus, except for an occasional attempt by a small company, there are no strong commercial applications of ethylene-removing films in the United States. On the other hand, sachets of potassium permanganate ethylene absorbers are in widespread commercial use.

BIBLIOGRAPHY

Abe, K. and A. E. Watada. 1991. "Ethylene Absorbent to Maintain Quality of Lightly Processed Fruits and Vegetables." *Journal of Food Science,* 56(6):1589–1592.

Anonymous 1990. "Zeolite Additive Keeps Food Fresher." *High Performance Plastic,* August.

Choi, S. O. 1991 "Orega Ultra-High Gas Permeability Film for Fresh Product Packaging," *CAP 91, International Conference on Controlled/Modified Atmosphere/ Vacuum Packaging,* Schotland Business Research, Inc , Princeton, New Jersey

Lasris, Barbara. 1990. "How to Get Rid of the Fox in the Hen House with Ethylene Scrubbers." *CAP '90 International Conference on Controlled/Modified Atmosphere/Vacuum Packaging, 1990 Conference Proceedings,* Schotland Business Research, Inc., Princeton, New Jersey.

Matsui, M. 1989. "Film for keeping freshness of vegetables and fruits." U.S. Patent 4,847,145, July 11.

Morita, Yoshikazu and Yuichi Fuji. 1992. "Method and apparatus for applying freshness keeping agent to a food packaging body." U.S. Patent 5,084,290. January 28.

Nakamura, Hachiro and Kiyotaka Omote. 1983. "Foodstuff freshness keeping agents." U.S. Patent 4,384,972. May 24.

Risse, L. and W. R Miller. 1986. "Individual Film Wrapping of Fresh Florida Cucumbers, Eggplant, Peppers, and Tomatoes for Extending Shelf Life." *J. Plastic Film Sheeting,* 2(2):163–171. April 2.

Rooney, Michael (ed.). 1995. *Active Food Packaging.* Glasgow, UK, Blackie Academic & Professional.

Szikla, Zoltan. 1993. "Ethylene Absorbing Paper for the Packaging of Fresh Fruits and Vegetables." Presented at *MAPack 93, Conference on Modified Atmosphere Packaging,* Greenville, South Carolina.

CHAPTER 8

Odor Removers

THE removal of odors from the interior of food packages may be both beneficial and detrimental. In the latter case, aroma capture may remove desirable components of the contained food, as when some interior polyolefin package components "scalp" aromatic notes from contained product contents such as orange juice. One function of the package, therefore, is to block the loss of desirable odor from the product. This is also often the primary objective of barrier packaging. In some instances, the aroma-removal properties are unintentionally incorporated into active packaging technologies, a topic discussed below.

The issue of the benefits of odor/aroma removal is, however, significant in the realm of active packaging. Many foods such as fresh poultry and cereal products develop what are referred to as confinement odors. Very slight and generally insignificant but nevertheless detectable deterioration odors, such as sulfurous compounds from protein/amino acid breakdown or aldehyde/ketone compounds from lipid oxidation or anaerobic glycolysis, may form during product distribution. These odors are trapped within gas-barrier packaging so that, when the package is opened, they are released to be detected by consumers. These relatively harmless odors, which generally do not signal any significant spoilage, may be cause for rejection even though they dissipate into the surrounding air within seconds. One reason for odor removal from the interior of packages would be to obviate the potentially adverse effects of these "confinement odors."

A second reason for incorporating odor removers into packages is to obviate the effects of odors developed in the package materials themselves. During plastic processing, i.e., extrusion, molding, film and sheet blowing, or casting, some polyolefin components may tend to break down or oxidize into short-chain and often odorous hydrocarbon compounds. Antioxidants are often included in the polyolefin-processing additive package to try to minimize the undesirable odor effects.

As another means of obviating the effect, DuPont has commercialized the incorporation of molecular sieves into polymer blends. Additionally, DuPont has patented other additives for incorporation into plastic packages whose purpose is to scavenge odors produced by food deterioration. The theory is that lipid-containing foods may oxidize into short-chain volatiles with undesirable odors. By removing these odorous compounds, the food quality is maintained at least perceptually without defects. Alternatively, removing the signals of food spoilage, as this technology does, may be counter to the idea that spoiled foods should not be consumed. Obvious questions include: When is food spoiled? When is food biochemically altered?

This chapter addresses various commercial and other technologies proposed to remove odors and aromas from food packages through the application of active packaging.

ACTIVATED CARBON ODOR ADSORBERS

Activated carbon is a popular odor adsorbent because it is very effective and inexpensive. However, it is black in color, picks up moisture, and, when heated or saturated with odor, can surrender some or all of its adsorbed odor.

A number of patents discuss odor removal by enhanced plastic- and paper-based materials. A U.S. patent assigned to Westvaco (Parks, 1996) describes a paperboard substrate extrusion-coated on both sides with low-density polyethylene (LDPE) A multiply layer is then coextruded onto the side of the coated substrate to contact the packaged product. This layer includes an odor adsorber such as activated carbon. In one manifestation, one surface of a paperboard substrate is coated with a 30% to 45% mixture of activated carbon in an aqueous binder system, applied at a coating weight of 3 to 26 lb. per 3,000 sq. ft. ream. The solids include 20% to 95% of activated carbon and 5% to 80% binder. The binder can be composed of sodium silicate, a polyester, starch, or polyvinyl alcohol. The coated paperboard is used to form folding cartons for packaging food products that emit "significant" odors or aromas. The coating, arranged on the interior side of the paperboard carton-forming board, is claimed to adsorb such odors.

ABSORBENT PAD ODOR REMOVERS

Numerous patents describe the removal of water and dissolved flavor components by absorbers in porous pads, usually applied to such products as diapers, but with potential to ultimately transfer the technology to packaging.

A Canadian patent (Gross, 1993) describes a method for reducing malodors arising during the use of a disposable absorbent product such as a diaper. This patent involves applying an effective amount of a surfactant, such as sorbitan monooleate, which is effective in reducing the odor of urine. The absorbent includes a water-swellable absorbent material. Surfactant is applied at 0.005 to 25 wt.% of the water-swellable material. One concept is that the surfactant may be applied to a pad within the package, or that it may be applied to one layer of a surrounding package structure to function actively.

Another Canadian patent (Tanzer et al., 1994) describes an absorbent pad that includes an anhydrous deodorizing non-buffer mixture of basic and pH neutral odor-absorbing particles, such as sodium carbonate and zeolite.

According to another Canadian patent (Yeo et al., 1991), an odor-absorbent covering for an absorbent pad may be produced by coating a nonwoven fabric with an aqueous composition incorporating a fluorocarbon polymer and an odor-absorbing reagent, such as sodium carbonate, potassium bisulfate, or citric acid; curing to form a hydrophobic coating around the reagent and to bind the reagent to the fabric; and forming 6 to 1,100 apertures per square inch in the fabric through which the odor may pass into the odor absorbent.

In a Japanese patent (Takada and Yasuda, 1993), a coated sheet adsorber contains a transition metal salt, such as titanium chloride (10 to 95 moles) and colloidal silica (10 to 90 moles) in water. Aqueous ammonia is added to the mixture to adjust the pH value to higher than 7. The co-precipitate of transition metal oxide and carrier is collected and burned at 200 to 800°C. The resultant material has a specific surface area greater than 100 m^2/g and is used primarily as a pigment for paper coating. The coated sheet is reported, however, to also adsorb unpleasant odors.

An odor-absorbing sheet material is prepared by bonding an air-permeable flexible cover sheet to a flexible substrate sheet, with odor-absorbing granules affixed in discrete areas between them. Sheet materials include fibrous, nonwoven, air-permeable polymer webs bonded with latex adhesive. The odor-absorbing granules may be activated clay, alumina, silica gel, odor-absorbing polymers, or activated carbon. The flexible sheet is claimed to be useful in applications such as removing odors in empty food storage containers.

A Canadian patent (Weber et al., 1990) describes a method for producing porous, non-dusting web materials such as sheets for absorbing odorous substances and for distributing absorbents such as zeolites or alkaline salts. This material is used in personal care products such as sanitary napkins, tampons, diapers, and incontinence pads. The material is also used in sealed packages containing medical supplies such as hospital gowns, whose offensive odors, especially acidic odors, are absorbed.

In an odor-retentive paperboard, lignin is the odor-removing agent. Lignin is added to the pulp stock in the beater of the paperboard-making machine and is also applied to the finished paperboard as a coating Paperboard treated in this manner is said to be suitable for use in food packaging. It absorbs and retains odors caused by printing inks applied to the paperboard, thus preventing these odors from being absorbed by the contained food.

In a Japanese patent (Itoh and Mori, 1991), an odor-absorbing corrugated fiberboard may be produced from 10% to 45% pulp slurried in water to which 60% to 85% aluminum hydroxide, 2% to 3.5% polyalkyl acrylate, and 0.9% to 1.5% polyalkyl acrylate are added. This mixture is converted to paper. Corrugated fiberboard is coated with an adhesive such as polyalkyl acrylate and glued between two layers of the paper. The resulting sheet is claimed to absorb unpleasant odors when used to wrap fresh foods.

In a Japanese Daiwa Spinning patent (Tokai et al., 1988), cellulosic fibers were treated with aqueous dispersions containing 0.5 to 2.5 N. NaOH or KOH and 0.5 to 50 g/L basic cupric carbonate. Rayon staple was treated with a liquid containing copper hydroxide and sodium hydroxide to produce cupric carbonate. The fibers with 12.5% copper content are claimed to possess properties of good adsorption of odors of inorganic or organic gases, amines, mercaptans, and aldehydes. Such an odor-adsorbing compound would not be accepted by U.S. regulatory authorities.

A Japanese Mitsui patent (Nakajima et al., 1984) describes aluminum hydroxide or oxide reacted with phosphoric acid in a two-phase system of water and an organic solvent to form fine crystalline ortho aluminum phosphate. The precipitated ortho aluminum phosphate is added to slurried pulp, which is converted to a paper sheet that can absorb unpleasant odors.

In a Japanese patent (Tsuchiya et al., 1980) zeolite powder (1.5%) is mixed with cellulose pulp in water. After dilution with water, the slurry is converted to paper, which can be used as an odor or moisture absorbent or filter.

Japanese tea manufacturer Hishidai Seicha has developed an antimicrobial and deodorizing sheet made from tea leaves. Tea leaves contain catechin, which is reported to be a highly effective deodorizing agent. The company's product, called "Ocha de Shinrin Yoku," is to be produced in conjunction with Tokai Pulp, which has developed a method to convert the components into a sheet to which chitosan has been added as an odor remover.

VITAMIN E AND BHT AS ODOR REDUCERS IN PLASTIC PACKAGING

Vitamin E, chemically alpha tocopherol, has been aggressively marketed as a food-grade odor remover in package materials. According to studies sponsored by vitamin E manufacturer Roche at Michigan State University School of Packaging, vitamin E can be a viable alternative to butylated hydroxytoluene (BHT) as an antioxidant additive in package materials. Antioxidants extend product shelf life by reacting with lipid peroxides or free radicals in dry fat-containing food products such as snacks, crackers, and cereals. Antioxidant additives also can help prevent plastic package material degradation during film blowing or bottle blowing, improve heat resistance, control color change during film extrusion or blow molding, and reduce adverse odor generation. Vitamin E has been used for a decade as an antioxidant processing aid by the plastics processing industry (Laermer et al., 1994).

The Michigan State University study compared three multilayer adhesive-laminated films, each an ionomer/ethylene vinyl acetate copolymer coextrusion sealant/high-density polyethylene (HDPE) lamination. The three films were identical except for the antioxidant incorporated into the core HDPE layer. One incorporated alpha tocopherol, one added BHT, and the control contained no antioxidant (Newcorn, 1997).

One series of experiments measured the time the antioxidants required to migrate into air at ambient and high temperatures from film samples, because the slower the migration, the longer the antioxidants remain in the film. The results indicated that alpha tocopherol consistently remained in the package film about ten times longer than BHT.

Another series of tests measured actual production shelf life. The film was fabricated into test pouches, each filled with oat cereal. Pouches were stored at ambient conditions for nearly a year. Evaluations measured the amount of lipid oxidation in the cereal and the amount of antioxidant remaining in the package material. No significant differences were detected in the level of oxidation in the cereal in test pouches made from the two antioxidant films. Packages using alpha tocopherol performed as well as those using BHT. Cereal from the control pouch without an antioxidant showed a marked increase in oxidation about halfway through the test.

After four weeks, no detectable levels of BHT remained in pouches made from the BHT-impregnated film. Further tests showed that 15% of the BHT had migrated into the product, continuing to provide a degree of protection against oxidation. By comparison, after 28 weeks, almost 30% of the alpha tocopherol remained in the packaging film.

Michigan State University researchers concluded that alpha tocopherol should be considered for incorporation into package materials for food products such as crackers or potato chips in which lipid oxidation is a major concern.

The effects of vitamin E and BHT as antioxidants on off-odor and off-flavor release from blow-molded HDPE bottles have also been compared. The effects of blow molding on the release of volatiles have been studied. Vitamin E was shown to be a superior antioxidant to BHT for reducing off-odor and off-taste.

MOLECULAR SIEVES AS PLASTIC ODOR REDUCERS

UOP Corporation reported on the characteristics and odor-adsorbing properties of a molecular-sieve technology, "Smellrite"/"Abscents." This material, a crystalline zeolite, has molecular sized organophilic openings or pores that attract and trap odor within its structure, lowering the odor concentration in the air to a level below the olfactory threshold. UOP's molecular sieves can be used as an odor-adsorbing material in incontinence and feminine hygiene products—as well as in active food packaging (Marcus, 1990).

DuPont has developed a series of molecular sieve compounds called OTC (odor and taste control), which reduce the flavors resulting from processing thermoplastic materials. These materials are different from scavengers incorporated into plastic materials, which are intended to remove those off-flavors arising from biochemical deterioration of some food products, especially fatty foods such as fried snacks (e.g., potato chips). The plastic odor-reduction materials are silicate-based molecular sieves with pore diameters of at least 5.5 nm. These materials absorb the off-odors resulting from plastic processing such as those generated by the heat of extrusion. An interesting finding is that the molecular sieves also apparently improve the hot tack, seal strength, and adhesion of ionomer materials to aluminum foil.

A paper/aluminum foil/polyethylene package for Dutch candies was the first application for "Conpol 20LO," an odor-absorbing additive by DuPont's Packaging and Industrial Polymers division in the United Kingdom. The additive, a molecular sieve containing silica, absorbs undesirable odorous molecules like aldehydes and ketones formed during processing of polyolefin sealants. The sealant contains 2% to 5% 20LO. The package is used by Van Melle, Breda, Netherlands, to contain its Meller chocolate and caramel sweets

ALDEHYDE SCAVENGERS

DuPont's studies on aldehyde scavengers have led to a series of patents (Brodie and Visioli, 1994). DuPont's initial development work was mostly aimed at removing undesirable components from the interior of the package. Some of the end-products to be removed are aldehydes, free fatty acids, free radicals, ketones, hydrogen sulfide, and polyphenol by-products. The end-products can be present in the headspaces of packages of snack foods, crackers and cookies, cereal, pet foods, rice, powdered dairy products, cooking oils, coffee, and soaps.

DuPont's initial investigations were with scavenging aldehydes such as hexanal and heptanal, which are formed from oxidation of fats and oils. These aldehydes can cause off-flavor and tend to impart a rancid taste to food products. A masterbatch containing the aldehyde scavenger can be incorporated into a number of different flexible package structures produced by converters. The aldehyde-scavenger-containing plastic could be incorporated into either non-barrier or barrier coextrusions for food products such as snacks, crackers, cookies, and cereals. The scavenger could be incorporated into a small pouch containing a coupon or sheet that would then be inserted into a larger bag; into the actual coupon; into a membrane lid between the barrier layer and the seal layer; or into the plastic used in a cap liner. It could even be used in one of the layers of a multilayer plastic container.

Some laboratory tests on aldehydes in the headspace were run on actual food products after exposing the packaged products to a temperature of 46°C (115°F). Studies on peanut butter packages comparing no scavenger in the lid to two concentrations of aldehyde scavenger demonstrated the effectiveness of the scavenger, particularly at the higher level, in removing aldehydes from the headspace. The scavenger was reportedly effective for powdered coffee lightener. Both peanut butter and powdered coffee lightener have normal shelf lives of up to nine months.

Results of testing for hexanal in the headspace of a snack product showed significantly less hexanal in the headspace of the bag containing the scavenger at the end of the product's normal shelf life (of several months), and the hexanal in the headspace was only half that of the standard package after it was exposed for twice the time. Aldehyde levels in instant coffee after 12 weeks of exposure were significantly reduced in total aldehydes with two concentrations of the aldehyde scavenger.

Interest was also expressed in scavenging sulfides found in packages of protein-containing foods and particularly in poultry. Laboratory tests showed that scavengers function in the removal of hydrogen sulfide (H_2S). DuPont injected a high level of hydrogen sulfide (12,000 ppm) into a vial containing two different types of scavengers and two different levels of scavenger. The results show that both scavengers were effective, with scavenger "A" being the more effective with the lower level of scavenger. Both scavengers were evaluated because the "A" sample would be about four times the cost of the "B" sample, and the "B" sample may be effective enough for removing lower levels of hydrogen sulfide. The sulfide scavenger could be incorporated into the dome lid of packaged processed cured poultry products. A sulfide scavenger could be incorporated into a label that would be placed inside the package.

Tests were conducted to determine if incorporating ultraviolet inhibitors into film would have a positive impact on retarding food discoloration. This testing was performed by compounding ultraviolet inhibitors into film, preparing bags from the film, and filling the bags with sliced pepperoni. Samples of two different inhibitors and sample bags without any inhibitor were aged under lights at room temperature, and the pepperoni within the bags was visually inspected over a seven-day period for color change. These results show that incorporating ultraviolet inhibitors into film significantly reduced food discoloration.

Work has also been performed to determine if the growth of mold in packaged cheese can be inhibited with the use of an aldehyde scavenger. DuPont researchers incorporated a number of different mold inhibitors and an aldehyde scavenger into films, prepared pouches from the films, and then filled the bags with shredded natural cheese. The pouches containing the cheese were then aged under light at room temperature and visually inspected for mold growth over a period of 23 days. The test samples were compared to a control pouch that did not contain an inhibitor. The results of mold growth versus time showed that the control sample developed mold growth over about 12% of the surface of the cheese, whereas the pouch with inhibitor "A" had mold on about 3% of the cheese. However, all of the other inhibitors tested prevented mold growth altogether. These results indicate that mold inhibitors incorporated into films can be effective in preventing mold growth.

Scavengers for the quinones from polyphenolic compounds were evaluated to determine if nonenzymatic phenolic browning of fruits could be reduced. This could lead to applications such as squeezable jam, jelly, and preserve containers. This testing involved testing jelly in glass jars with "coupons" of scavenger film on the bottom, exposing the jars to 50°C, and then measuring the ultraviolet absorbance. Ultraviolet absorbance was measured at 440 nm, which relates to browning and pigment degradation, and at 500 nm, which relates to overall color intensity. The results show that the scavenger permitted less color change than the control.

One of the basic DuPont patents offers some insights into the nature of the scavenger family. DuPont's materials are compositions of polyalkylene imine (PAI), particularly polyethylene imine (PEI), and polyolefin polymer (including copolymers) to produce package films for oil-containing foods. The film is capable of scavenging unwanted aldehydes from the food product. More specifically, the preferred compositions of the invention comprise a discontinuous PAI phase and an olefinic polymer continuous phase in a weight ratio of PAI to olefinic polymer of about 0.001 to 30:100. The polymer is extruded into film or sheet. Claimed is a PAI having the structure:

$$H-(CH_2-(CH_2)n-NH-)m-H$$

where m is at least 1 and n is in the range from 1 to about 4, and the molecular weight is greater than about 800. The olefinic polymer can be a very wide variety of polymeric materials, including ethylene vinyl acetate, polypropylene, polyethylene, and ethylene vinyl ester copolymers.

It is important that the olefinic polymer have properties sufficient to allow the polymer to be converted into a thin film or sheet for use in packaging applications. The olefinic polymer should be capable of heat-sealing to itself in a packaging application. The combination can be incorporated into a film and placed in direct or indirect contact with a food product. The film can scavenge unwanted aldehydes from the food product but will generally not allow PAI migration from the film.

Olefinic polymers include homopolymers such as polypropylene, low density polyethylene (LDPE), linear low density polyethylene (LLDPE), and ultra low density polyethylene (ULDPE). Critical to the olefinic polymer is that it be capable of being formed into a thin sheet or film, such as for a conventional packaging application. In an alternative embodiment, the olefinic polymer is heat-sealable to itself.

The preferred PAIs contain a high percentage of nitrogen. The patent states avoidance of very-low-molecular-weight materials that can migrate. The molecular weight should be above about 800, more preferably above about 1500, and most preferably above about 2500.

The final film composition was found to be stable. Migration or surface blooming of the liquid PAI within the olefinic polymer was not found to be a problem. By incorporating the PAI into an olefinic polymer, the PAI can be incorporated into an existing layer of a conventional multilayer packaging. The PAI was able to maintain its aldehyde-scavenging properties, even when alloyed into an olefinic polymer. Although carboxyl groups tend to react with the PAI and so diminish the efficacy of the PAI, most olefinic polymers offer a matrix on which PAI can be incorporated.

A functional barrier is placed over the PAI/olefinic polymer film. The functional barrier should be permeable to aldehydes but impermeable to PAIs. Generally, the functional barrier can be extremely thin, since the PAI is already locked into the olefinic polymer matrix. Generally, the film can be as thin as possible without allowing for holes or nonuniform coverage. Functional barriers of 0.003 inch or greater gauge are possible, although thicknesses of less than 0.001 inch are preferred. The preferred functional barrier layer is an olefinic polymer, preferably an olefinic copolymer derived from an olefin monomer, such as ethylene and one or more vinyl ester (or acid derivatives) comonomers. Examples of useful functional barriers include ethylene vinyl acetate, ethylene (meth)acrylic acid, or blends of these or other olefinic polymers. In the preferred embodiment, the functional barrier is capable of heat-sealing to itself. Alternatively, the composition contains a binding agent that would allow the composition to be used as the food contact and/or heat-seal layer.

Although the scavengers will not stop the oxidation reactions, they will remove the end-products and thus eliminate any sensory signals that would otherwise lead to rejection of what might be good food.

AMINE ODOR REMOVAL

Amines formed during fish muscle degradation include strongly basic compounds which are potentially strong in their interaction with acidic compounds such as citric or other food acids. Incorporating these acids in polymers, such as polyethylene, and extruding them as layers in packaging have been demonstrated to remove amines.

A more recent approach to removal of amine odor has been provided by Japan's Anico, under the tradename Anico Bag. Bags made of film containing ferrous salt and an organic acid, such as citric or ascorbic acid, are claimed to oxidize the amine or other oxidizable compound as it is absorbed by the polymer (Rooney, 1995).

CONCLUSION

Odor removal has been the subject of considerable research. Diaper odors have been one particular targeted topic, with numerous mechanisms such as the incorporation of activated carbon, fluorocarbons, sodium carbonate, alkaline salts, and molecular sieves as some alternative odor removal technologies.

Vitamin E has been proposed as a mechanism to reduce odor production, especially as a butylated hydroxytoluene replacement in materials for packaging snacks and hard bakery goods.

DuPont has performed extensive research on incorporating aldehyde scavengers into package materials to remove the end-products of lipid oxidation. Among the possible alternatives might be an additive effect of vitamin E and aldehyde scavengers, with the former delaying oxidation and the latter removing the odorous compounds, thus effectively significantly prolonging the shelf lives of contained foods.

BIBLIOGRAPHY

Adcock, L. H. 1968. "Producing odor-retentive paper, board, and like materials." Canadian Patent 775,023. January 2.

Brodie, Vincent. 1994. "The New 'In-Package' Scavenger Technology and Its Probable Impact on Package Structures." Presented at *Institute of Packaging Professionals Packaging Technology Conference* and published in its proceedings volume November 11

Brodie, Vincent and Donna L. Visioli. 1994. "Aldehyde scavenging compositions and methods relating thereto." U S. Patent 5,284,892. February 8

Gross, J. R. 1993. "Method for reducing malodor in absorbent products and products formed thereby." Canadian Patent 2,072,914. September 28.

Itoh, O. and H. Mori. 1991. "Deodorant sheet." Japanese Patent Kokai 109,919/91. May 9

Laermer, S F., S. S. Young, and P. F. Zambetti. 1994. "Could Your Packaging Use a Dose of Vitamin E?" *Food Processing*. June.

Laermer, Stuart F, Sai S Young, and Peter F. Zambetti II. 1996. "Vitamin E Cuts Plastic Taste in Flexible Food Packaging." *Converting,* February.

Marcus, B. 1990. "Smellrite/Abscents—Unique Additive for Odor Control." *TAPPI Nonwovens Conference,* Marco Island, Florida, Proc.: 283–291. May 6–10.

Nakajima, Y., A. Tsuwako, K Maruyama, S Ihno, and M. Takenaga. 1984. "Paper for absorbing unpleasant odors." Japanese Patent Kokai 95,931/84. June 2.

Newcorn, David. 1997. "Not Just for Breakfast Anymore" *Packaging World,* 4(3), March.

Parks, C. J. 1996. "Odor-sorbing packages." U.S Patent 5,540,916. July 30.

Rooney, Michael. 1995. *Active Food Packaging.* Glasglow, UK: Blackie Academic and Professional.

Shrivastava, A , A. L. Vainshelboim, and J. R. Wagner 1996. "Selective Absorbent of Ammonia with Biological Activity for Odor Control of Disposable Diaper Products" *1996 Nonwovens Conference: Proceedings,* TAPPI: 277–282, TAPPI Press. March 11.

Takada, S. and T. Yasuda. 1993. "Highly adsorptive pigment" Japanese Patent Kokai 214,263/93. August 24.

Tanzer, R W., M. A. Brummer, and A. A. Gossens. 1994. "Absorbent article containing an anhydrous deodorant." Canadian Patent 1,328,987. May 3.

Tokai, Yu, M. Ootaguro, and Y. Komatsu. 1988 "Manufacture of odor-absorbing cellulosic fibers." Japanese Patent Kokai 235,571/88. September 30

Tsuchiya, K., S. Shimaguchi, and K. Honda. 1980. "Absorbent paper" Japanese Patent Kokai 62,299/80. May 10.

Uegakito, O., K. Hayashi, M. Sugiura, K. Fukushima, and M. Horii 1986 "Odor absorbent." Japanese Patent Kokai 136,438/86. June 24

Weber, M. G., S. W. Fitting, R. E. Weber, and R. S Yeo. 1990. "Odor-absorbing web material and method of making the same and catamenial devices and medical material packages containing web material." Canadian Patent 2,014,204. October 14.

Yeo, R. S , M. G. Weber, S. R. Majors, and R. W. Tanzer. 1991. "Odor-removing cover for absorbent pads and method of making same." Canadian Patent 2,043,434. December 21.

CHAPTER 9

Aroma Emissions from Plastics

CONTROLLED release of desirable aromas from plastic package materials has been suggested as a means to enhance the flavor perception of contained food on opening and shortly thereafter. The volatility of an aromatic material is vital to its performance in a flavor and especially in release applications. The volatility of aroma chemicals is largely dependent on their molecular weights, which range upward from the simple two-carbon structures of acetaldehyde and acetic acid. The majority of useful materials range in molecular length from 6 to 18 carbons, with those over 20 carbons usually exhibiting too low a volatility to be useful as aroma producers.

Flavors may interact with plastics by two distinct pathways. The first is a function of the aroma compound's chemical compatibility or solubility with the plastic package material. The second is the more subtle diffusion into or permeation through the plastic package material by components of flavors. In most cases of odor/flavor scalping, which is basic to aroma emission from plastic package materials, the mechanism responsible is a combination of the two.

Chemical solubility depends on organic functionalities of the plastics and the aroma compound, the molecular weight of plastic material, and the degree of crystallinity or orientation of the plastic. Chemical solubility decreases as the plastic functionalities and the aroma chemicals become more dissimilar, as the plastic's molecular weight increases, and as the degree of crystallinity or orientation increases. It is rare to find a high degree of chemical solubility in packaged foodstuffs because of the dilution of aggressive organics in heavy oils or aqueous media.

Flavors diffuse into or permeate through plastic by virtue of the free volume and porosity of the polymer. Plastic materials contain unoccupied space or free volume within their polymer matrix As the free volume increases, the mobility of the polymer chain segments increases, and the diffusion of small volatile molecules increases. The factors affecting absorption of an aroma into a plastic's free volume include the size and flexibility of the chemical, the volatility of the aroma chemical, the plastic material's thickness, and factors affecting polymer chain mobility such as degree of orientation and degree of crystallinity. Polymer porosity results from voids of various sizes forming networks of interconnecting channels on a much larger scale than free volume. The formation of these voids depends on the processing conditions and additives incorporated into the plastics. The factors affecting absorption resulting from a plastic material's porosity are much the same as free volume, but are not as dependent on the size of the flavor chemicals.

The absorption of aromas into plastics and hence the release of aromas from plastics is basically an equilibrium process. The chemicals move from a site of high concentration in the product to a site of low concentration in the plastics. Increasing the concentration of the flavor chemicals being absorbed could increase the problem of flavor scalping rather than solve it. Flavor in this context is really aroma since the taste component of the flavor is usually immobile and not volatile, and therefore cannot be "scalped." In addition, storage at elevated temperatures could shift the equilibrium to increase absorption into the plastic.

Research has been conducted on means of introducing aroma into plastics to enhance the plastic functionality. Flavor supply companies have developed concentrates and processing techniques to disperse aroma compounds in conventional plastics through processes such as extrusion. Polymer concentrations with high load levels of flavor can provide a cost-effective method of masking nuisance malodors. These same concentrates can be blended into food package plastics to eliminate or significantly reduce flavor scalping by removing the driving force from the equilibrium process. Flavor concentrates might impart an added value to packages by improving aroma on opening or by attracting consumers with aroma-emitting package materials.

Several U.S. plastics-compounding companies market the following:

- flavor/polymer alloys loaded with flavor concentrations from 25% to 40%
- free-flowing pellets
- thermal "protection" of aroma under conventional plastics-processing temperatures
- controlled release of the aroma
- compatibility with a wide variety of base plastics

Some materials are available for direct use in polyethylene, polypropylene, ethylene vinyl acetate, ionomer, nylon, polyester, and polyvinyl chloride plastic materials. The plastics processes in which the system is claimed to be able to be used include injection molding, film blowing and casting, sheet extrusion, vinyl calendering, foaming, hot-melt adhesives, and hot-melt coating.

POLYVEL

Polyvel Inc , (Hammonton, NJ), a U.S. flavor concentrate compounding company, claims to have developed the ability to incorporate fragrance or aroma into plastic as a result of expertise in the development of high-end additive concentrates for the plastics industry. Specifically, Polyvel specializes in concentrates made by incorporating liquid additives, such as fragrance oil, during extrusion With polymer chemistry experience, Polyvel has developed what it believes to be a high-quality line of fragrance concentrates and foamable fragrance plastic pellets. The pellets contain gas emitters that may foam the plastic into a low-density material. They claim extensive experience with additive concentrate processing.

Polyvel fragrance concentrates for cosmetics are provided in extruded pellet or powder form and can be incorporated into a plastic molding or extrusion process by adding a solid, pelletized color concentrate. Since Polyvel concentrates are sold in pellet or powder form, they claim neither problems of sticking nor clumping, as might occur if liquids or damp pellets were used to ensure thorough distribution of fragrance concentrate in the masterbatch and feeder, maintaining consistent fragrance level and quality of finished plastic parts.

All Polyvel fragrance concentrates have been developed and are formulated for high-temperature plastics processing and molding applications, ensuring effective transfer of active fragrance to finished plastic parts.

Polyvel fragrance concentrates can be formulated to last from a range of days to years in the pellets, or they can be tailored as needed for particular applications.

Polyvel P25 fragrance concentrates contain 25% active fragrance oil in a polyethylene carrier resin. Polyvel recommends their P25 concentrates for use in polyethylene, polypropylene, and ethylene vinyl acetate plastics. Polyvel P25 fragrance concentrate can be used at low levels to impart fragrance to extruded and molded articles including sheet and film. Whether to mask an unpleasant odor resulting from plastics processing, incorporate fragrance into a product, or use in a new or existing air freshener, P25 is claimed by Polyvel to be a good choice for cost-effective fragrancing of polyethylene, polypropylene, or ethylene vinyl acetate. P25 concentrates are available in many fragrances and flavors and can be manufactured with custom fragrances.

The following are typical properties of P25 concentrates:

- concentrate form: dry pellet
- fragrance oil level: 25%
- softening point: 110°C
- melt index: 10 g/100 minutes

Of major concern about Polyvel are: (1) the thermal stability of their concentrates—more a question of degradation, particularly with sweet flavors such as chocolate, than with flavor loss; (2) dispersion in the carrier resin; and (3) uniform distribution in the end use.

The word *fragrance* may be regarded as marketing hyperbole when applied to garbage bags because the fragrance dissipates within minutes of removal of the bag from its secondary package. Film bags are made of film 0.002-inch or less gauge and contain only 0.25% to 0.35% fragrance. Polyvel recommends a venting system be installed near the film extruders to assist in clean-up and purge. Problems can occur if two blown film lines are running side-by-side, one with a fragrance concentrate and one without; there can be odor transfer to the unscented bubble. Also, in distribution, care needs to be taken to use barrier packaging (aluminum foil or nylon based) on "fragrant" rolls to retard odor transfer between rolls and to other materials.

Polyvel treats their flavor/aroma compound concentrates similarly to color concentrates. With the application of pigment, the operator can determine when one concentrate has been converted to another (or to a nonodorous compound). Another factor is that the fragrance concentrates do not tend to adhere to metals or to end up in extruder nooks and crannies.

In the Polyvel concentrate-compounding system, flavor oil is injected into the barrel of extruder. A metering or injection pump is calibrated with plastic resin feed. The extruder screw does the mixing to generate microbubbles of liquid, which is the key process. After extrusion, the resin is cut into pellets. Aromas in concentrated compound pellets will develop an equilibrium with headspace in the package containing the plastic resin. Lower temperatures reduce volatilization of the flavorant. Film cast at 270°C (518°F) can lose up to half of the active flavor incorporated. Film blowing can result in reduction of up to 15% of the aroma. In films, typically 0.2% active aroma ingredient is present.

FRAGRANCE CONCENTRATE SUPPLIERS

A major U.S. plastic concentrate compounder, Ampacet Corporation (Tarrytown, NY), supplies fragrance concentrates that are typically pellets in a polyethylene carrier resin together with a colorant. They are used in non-food applications such as bathroom and kitchen trash can liners. Fragrances include pine, lemon, and baby powder scents. Garbage bags produced by Armin, whose "Ruffies" products incorporate fragrances, can withstand polyethylene-processing temperatures (molding or extrusion or blow molding) (Table 10).

AROMA CONCENTRATES

Dragoco is a flavor supply company that does not offer any commercial aroma concentrates for plastics, but it was prominent in the past in developments relative to aroma release. Dragoco's approach involved adhering a lam-

TABLE 10. Garbage/Trash Bags with Scents in Retail Distribution, 1999.

Brand	Product Name	Supplier	Type Bag	Typical Scents	Outer Package
Ruffies®	Color Scents	Carlisle Plastics, Inc. (Armin), Minneapolis, Minnesota	Tall kitchen bags	• Citrus spice • Mountain heather • Fresh-cut flowers	Polyethylene overwrap
GoodSense®	Kitchen Scentsations	Webster Industries, Division of Chelsea Industries, Inc., GoodSense Division, Peabody, Massachusetts	Tall kitchen bags	• English garden • Cinnamon clove • Vanilla bean	Polyethylene overwrap
I-Deal®	Color Rite Design Scents	The Interplast Group, Consumer Division, Livingston, New Jersey	Tall kitchen bags	• Powder blue • Wildflower	Polyethylene overwrap
Hefty®	Basics Colors	Pactiv, Lake Forest, Illinois	Tall kitchen bags	• Spring bouquet scent	

Note: (1) all the bags are color-coordinated with scents, and (2) the bags are 13-gallon sizes.

ination to the outside of a package, e.g., a bottle. A tear strip would be attached to the closure. When the closure was removed from the bottle, the strip would cause the lamination to delaminate, thereby releasing the aroma.

The "inclusion" process provides for slow release of materials embedded in plastics. The process is used by Japan Liquid Crystal for food and drug applications. For example, use of this technology can release a deodorant in packaged fish or a sweetener in a "bad-tasting" drug. The process may be viewed as a means of releasing chemicals in permeable plastics. This technology appears especially good for adding volatiles, which are otherwise difficult to mix in plastics. Molded parts could include process control agents, fragrances, and even microbicides. The rate of release depends mostly on the degree of permeability of the host polymer.

The key to aroma inclusion is cyclodextrin, a modified starch that captures what is called the "guest" material. Starch/guest compound is offered as a masterbatch to be molded with a permeable plastic. After the part is molded, the guest material is gradually released.

The modified starch inclusion material is heat-sensitive, limiting its use to a processing range of about 140°C to 230°C (284°F to 446°F). Further, some of the guest material, i.e., aroma emission, is lost during molding (Anonymous, 1987a).

EXTRUSION BOUNDARIES

Contrary to the claims by flavor and concentrate supply organizations, plastic film converters involved in the incorporation of aroma compounds into their products have reported that the aromas arising from volatilization during extrusion tend to permeate the entire plant. Thus, it is necessary to isolate the film extruder to avoid contaminating the entire manufacturing operation with aromas that may be undesirable for most applications. Cast film extrusion lines operate at temperatures too high to permit them to be used for production of film that incorporates aroma. Blown film extrusion lines permit the incorporation of aromas at the very lowest levels required to permit later emission. Such flavors from concentrates may be successfully incorporated into polyethylene films. Once the film is in tightly wound roll form, the aroma is effectively trapped so it is not lost and does not contaminate other package materials. However, during packaging operations, such as on vertical form/fill/seal machines, the aroma is emitted and can probably contaminate nearby products, package materials, equipment, and operators. Of particular concern are the seal areas where the heat can vaporize aromas.

One means to eliminate the problem would be to trap the aroma between layers of a coextrusion with an odor barrier on the exterior, thus dictating a significant increase in package material cost. The aroma layer would be on the interior in contact with the food contents only.

Among the concerns surrounding this concept are the changes in flavor of the product, which arise from the incorporation of aroma into the surrounding package materials.

OTHER SUPPLIERS/TECHNOLOGIES

Plastiflac, a Belgian injection and extrusion blow molder, produces a range of plastic closures, bottles, dispensers, etc. A most interesting offering, however, is a scent-emitting injection-molded plastic piece. Plastiflac apparently injects appropriate scents into the plastic melt, which is injection molded, and then the fragrance is emitted in the region surrounding the piece. No indication has been offered as to the time for scent emission or to the limitations of the types of compounds that could be incorporated.

A small U.S. entrepreneurial company, Vista International Packaging, has developed sausage casings that are coated on the interior with spices and/or seasonings. Rather than blend the flavors into the meat emulsion or place the spices on the sausage surface where they might fall off, the casing contains the flavoring that is imparted to the sausage contained in the casing. In this manner no hard exterior rind is created.

Large-pore silica gel can be loaded very slowly with liquid aroma. The aroma is then emitted with the rate of emission higher at higher temperatures or with water.

The decoration and finishing of printed packages with scented lacquers and film or foil laminates requires careful preparation, notably thorough drying of the print layer. Solvent- and alkali-resistant inks are needed for safe and durable color reproduction. Any print powders should be based on starch and distributed finely and uniformly. Surface tension between substrate and ink must not fall below 35 mN/m. Lacquer choices include numerous dispersion formulations with glossy, matte, nacreous, barrier, skin, blister, and other effects; scented lacquers; primers; and solvent-based and radiation-hardening formulations. Laminating materials under consideration include polypropylene, soft polyvinyl chloride, polyethylene terephthalate, embossed polyethylene, and Okophan (a starch-containing biodegradable polyolefin) films; aluminum foil; film/paperboard, film/aluminum foil, and other flexible laminations; and plastic (coated polypropylene or polyethylene terephthalate) windows for folding carbon display packs.

Japan's Kuraray introduced plastic films containing fragrances. The films are heat-sealable ethylene vinyl alcohol (EVAL®) EF-HS. Test results comparing them with other films for flavor absorption and showing distribution ratios of volatile compounds between ethylene vinyl alcohol films and sample solutions indicated effectiveness. For market success, a flavor-retaining film must have good gas-barrier properties, certainly a characteristic of EVAL, and be nonabsorbent

Fragrance can be imparted by direct impregnation with liquids, printing with scented ink, deposition of porous adsorbent mixed with binder, silk-screen printing with microencapsulated fragrance, and lamination with scented film. Japan's Kohjin Company and Nippon Petrochemicals Company claim to be the world's only producers of multilayer paper-laminated scented plastic film. Applications range from pocket calendars to bookmarks to greeting cards. Easily stored peel-off-type laminations using barrier film are favored by cosmetics manufacturers for samplers and catalogs. Barrier-type two-layer lamination is engineered for tissue and lingerie packaging. Analogous products using other volatile chemicals include deodorizing film for waste bags, peel-off insect-repellent film, mosquito-repellent dog collars, and mold-proofing bags for leather garments.

Japan's Okamoto Industries Inc. is marketing a soft polyvinyl chloride "Aroma Film" series, which can emit different fragrances. Through special processing, "Aroma Film" provides 12 types of capsule-perfume fragrances such as lavender, strawberry, lemon, etc. Moreover, combinations of Aroma Film can be delivered in 16 color variations, allowing this product to expand its uses for bags, stationery, etc., by selecting fragrance, color, and pattern to match use and design. The fragrance of Aroma Film increases when the patterned parts into which fragrance has been applied are rubbed, thus allowing users to further enjoy its fragrance. The fragrance retention term is reportedly one year. Aroma Film is offered in four gauges: 0.3, 0.5, 0.8 and 1.0 mm.

In the United States, fragrances and flavors have now been applied as a coating over four- or six-color printing. Aromatic oils, entrapped in a hot-melt polymer adhesive system, are applied while the press is running. Substrates include paper, plastics, cloth, and ribbon. Cosmetic companies provide the essential oil for the fragrances; flavors such as chocolate are created in-house when advertisers seek the smell of their flavored products promoted in magazines. Creative Environments, Inc. developed and patented three polymer fragrance technologies. The first technology traps the fragrance in a polyamide resin and then releases the fragrance, an action that increases rapidly at temperatures of about 20°C (68°F). The second uses a nonflammable water-soluble resin; this process does not involve slowing the press, does not require a drier, and is applicable to newspaper web rotogravure. In magazine advertising, the aroma is not detected until the relevant page is opened.

Fragrant films tend to be poor in mechanical adaptability, short in fragrant life, and problematic in plant production environments. Apart from direct addition and cyclodextrin mixing, the fragrance can be dissolved in plastic material to produce a masterbatch that is then blended with plastic for fragrant film production. Barrier tape laminated with fragrant film can be folded for the fragrance to be diffused to an enclosed space and printed on an external surface. The fragrance can also be contained by peel-off barrier tape. The technology of fragrant film has been extended to deodorant, animal and bird deterrents, and moisture-prevention film for clothing storage.

Japan's Hexachemical Company succeeded in mixing their Celluresin cyclodextrin with added fragrances and insecticides with polyethylene and polypropylene without heating, allowing products to be manufactured with an insecticidal effect. Guest materials such as flavors/fragrances, insecticides, antimold agents, and deodorants are combined with cyclodextrin and kneaded with a synthetic resin to form pellet-like masterbatch for sheeting and injection moldings.

IFF, a major global flavor producer, suggests the use of odorants in plastics products such as refuse bags and other packaging items, hospital and personal hygiene products, synthetic leather and wood, etc. Such additives may also be used to mask unpleasant odors in plastics, e.g , those caused by styrene monomer and phenol in certain thermoset plastics. IFF has developed pelletized fragrance concentrates designated Polyiff, which can be batch-blended with polypropylene or polyethylene pellets. The company claims the pellets are easier to process than liquid additives and that any fragrance can be incorporated. Available scents include fruit, floral, and perfumes, at 1% to 5% typical use levels. Applications include disposal bags.

A Dow Corning/Felton Polytrap polymer-entrapment system uses a hydrophobic thermoset polymeric lattice network to hold a fragrance and permit continuous, controlled release. The Polytrap material consists of a colorless, nontoxic material added prior to polymerization. The resultant lattice network entraps volatile or migratory additives. The system has proved useful with fragrance oils and is being evaluated for antimicrobials, animal repellents, and slip agents. It is claimed to be usable in virtually any plastic resin, including polyvinyl chloride, polystyrene, and acrylonitrile butadiene styrene.

Wickhen Products has produced a plastic with fragrance built in to simulate fruit, etc., aroma. The system, Polytrap thermoplastic fragrance concentrate, allows high levels of fragrance to be added to most thermoplastics. It is available in powder, bead, flake, or pellet form as a multifunctional concentrate masterbatch.

The addition of fragrance additives to items such as plastic garbage bags and personal care items is increasing. Fragrance additives may be used to add an aroma to an article or to mask undesirable plastic odors. Fragrances result from the mixing of aroma chemicals, which may be synthetic or natural essential oils. These may be used in pellet or powder form for addition to thermoplastic resins during processing. The fragrance may be added in a straightforward melt blend, although some use a sponge-like microporous polymer or a thermoset resin matrix to trap the fragrance during addition to the melt. The release of fragrance from the finished article depends on the cross-sectional thickness of the plastic article, its surface-area-to-volume, the fragrance ingredients, and cost. Careful handling and storage of fragrance additives are recommended followed by careful purging of equipment after using the additive.

TechniChem Inc. has introduced a range of dry fragrance concentrates suitable for use in plastics compounding. The products are available in powder form, and the carriers used are claimed to be compatible with most resins. Effectiveness is claimed not to be lost under high-temperature processing. Typical fragrances include lemon, lime, rose, cherry, chocolate, and leather

Enka fragrance carriers are manufactured in a process in which a volatile aroma liquid chemical is incorporated in a microporous polymer matrix that serves as a secure storage medium. The fragrances reportedly do not change the physical properties such as tensile strength and elongation of polyethylene and ethylene vinyl acetate with varying concentrations of fragrance additives. The liquid aroma is said to be incorporated into a microporous polymer structure and then apportioned to the neutral granulate. The perfume is emitted by controlled release over a long period. (Anonymous, 1983a, 1983b, and 1983c)

The aroma is incorporated in a microporous polymer matrix. In the molding stage, fragrance carriers are added to the bulk granulate. The polymer matrix carrier protects the fragrance from degradation during heating and releases it into the bulk of the plastic only during the compression stage of the extrusion process. The scent is evenly dispersed throughout the finished product and is released in a controlled manner.

Using what they claim to be an innovative polymer-based technology, Janco Press, Inc. produces pressure-sensitive adhesive labels containing air-activated fragrances. The polymer-based system, which uses a product called Techscent, releases the fragrance when exposed to air and does not require customer interaction, a distinct advantage over traditional microencapsulation methods. Unlike encapsulation, the process does not alter either the applied fragrance or the printed product. Because the scent remains inactive until exposed to air, an applied fragrance can have a shelf life of months or years, depending on the display or storage method. Janco's production equipment includes a letterpress featuring in-line folding and gluing (Anonymous, 1990a).

In another Japanese technology, a paper is laminated with a thermoplastic sheet containing an odorized porous inorganic adsorbent to form a cup that emits fragrance when filled with hot water. Thus, paper is laminated with a polyethylene film containing a molecular sieve or a zeolite, sprayed with a tea essence, and molded into a cup (Anonymous, 1985a).

The Fretek wafer was introduced in 1989. Developed by a Japanese company and marketed in the United States by Techno International, this multilayer wafer is inserted into the package to alter the internal environment. The active ingredients are ethyl alcohol and acetic acid plus a flavor. The ethanol is an antimycotic agent and the acetic acid "enhances" the microbicidal effect. The wafer can also be formulated to bear an aroma similar to that of the product itself and thus compensate for aroma loss. The active ingredients are impregnated into a paper pulp layer sandwiched between two layers of perforated film. The technology is fundamentally an ethanol microbistatic agent to be inserted into a bakery goods package. The odor of the alcohol is masked by the added aroma.

BIBLIOGRAPHY

Anonymous 1982. "Sweet-Scented Plastics." *Manuf. Chem* , 53(12):25 December.

Anonymous. 1983a. "Enka Fragrance Carriers Give Plastics a New Dimension." *Obernberg*, 6.

Anonymous. 1983b "Easy To Use Fragrance Pellets Perk Up PE, PP." *Plastics World*, 41(6):88. May.

Anonymous. 1983c. "A New Invention from Enka Makes it Possible to Give Plastics a Permanent Fragrance." *Silk Screen*, 31(6):228–229 June

Anonymous. 1984a "Dry Fragrance Concentrates." *Modern Plastics*, 61(4):140. April.

Anonymous 1984b. "New Concentrate Technology for Fragrances and Other Additives." *Plastics Technology*, 30(4):12. April

Anonymous. 1984c. "Aroma Additives Are Effective, Economical" *Modern Plastics Int.*, 14(6):12. June.

Anonymous. 1985a. "Fragrance-emitting cups from paper laminates" Japanese Patent Kokai 27,538/85. February 12.

Anonymous. 1985b "Making Plastics Fragrant" *Kunststoffberater*, 30(7/8) 34–35. July/August.

Anonymous 1985c "Fragrance Release Rearranges Molecules" *Plastics World*, 43(9):98 August.

Anonymous 1987a "Cyclodextrin-Applied Plastic Products Commercialised." *Japan Chemical Week*, 28(1410):3 April.

Anonymous. 1987b. "Fragrances for Plastics." *Mod. Plast. Int.*, 17(5) 77 May.

Anonymous. 1987c. "Plastics Having Sweet Smell to be Sold" *Japan Chemical Week*, 28(1423)·3. July

Anonymous. 1989. "Unique Process Embeds Volatiles in Plastics" *Plastics World*, 47(1):58 January.

Anonymous. 1990a. "Scents Appeal" *American Printer*, 204(5):38–39 February.

Anonymous. 1990b. "Labels with That Sweet Smell of Success." *Converting Magazine*, 8(10):57–58, 60, 62, 64 October.

Anonymous. 1991. "Flavour Holding Resin" *Plast. Ind. News* (Japan), 37(10):146. October.

Anonymous. 2000. "New Technology Incorporates Fragrance into Polyethylene Resins." *Food and Drug Packaging*, 65(5):13.

Barth, P 1995. "Package Finishing with Lustrous Scents Laminating and Lacquering." *Coating*, 28(9):337–339, September

Booma, M., H. Hoojjat, and J. R. Giacin. 1993 "Permeability of Fragrance Volatiles Through an Ethylene-Vinyl Acetate Co-Polymer Membrane Based Delivery System: The Effect of Time and Temperature." Presented at *1993 Polymers, Laminations and Coatings Conference*, TAPPI (Chicago, Illinois), August 29–September 2, Atlanta, Georgia.

Cho, J. 1993. "Market Trends and Future Prospects for Fragrance Retaining Films." *Packpia*, 7:68–71.

Saito, H. 1989. "Plastic Film with Gaseous Chemicals for Multifunctional Paper." *Japan J. Paper Technology*, 7:18–21. July.

Saito, H. 1991. "Scented Paper." *Japan Journal of Paper Technology*, 2:144–145. February

Suran, V. 1986. "How to Use Fragrance Additives." *Packaging Review* (S. Africa) 12(2):59, 60–61. March–April.

Suran, V. M. 1986. "How to Use Scented Additives." *Macplas*, 11(77):70–71, April.

Tashiro, M., T. Motomiya, and K. Ori-i. 1992. "Synthetic paper for regulated release of aroma." Japanese Patent Kokai 263,609/92. September 18.

Venator, Thomas E. 1986. "Fragrance and Flavour Interactions with Plastics Packaging Materials." *Packaging*, 58(668):30–32. October.

CHAPTER 10

Antimicrobial Packaging

COMMERCIAL antimicrobial food package materials and structures still appear to be a few years in the future, but efforts of the 1990s underscored the continued development of new functional roles for packaging materials (Rice, 1995). Hotchkiss (1995) enumerated alternative antimicrobial packaging and summarized the issues relating to antimicrobials in package structures and materials.

In most, but not all, solid or semisolid foods, microbial growth occurs primarily at the surface. In prepared or mixed foods, microbiological growth can occur anywhere in the mass. Antimycotic agents are not uncommonly incorporated into waxes and other edible coatings used to package hard-skin fresh-produce items such as oranges and apples. More recently, the concept of incorporating antimicrobial agents directly into package films that contact the surface of the food has been developed, particularly in Japan, but also in other geographic regions.

Antimicrobial package materials may be classified into two types: those containing antimicrobial agents that migrate to the surface of the package material and thus can contact the food, and those that are effective against food-surface microbiological growth without migration of the active agent(s) to the food. Several commercial antimicrobial films have been introduced, principally in Japan. One is a synthetic zeolite that has had a portion of its sodium ions replaced with silver ions, which can be antimicrobial under certain situations. Such zeolite may be incorporated directly into a food-contact film. The purpose of the zeolite apparently is to allow for the slow release of silver ions to the food. Only a few scientific descriptions of the effectiveness of this material have appeared, and the regulatory status of the intentional addition of silver ions to foods through the medium of package contact has not been clarified by U.S. regulatory authorities.

131

Several other synthetic and naturally occurring compounds have been proposed and/or tested for antimicrobial activity in packaging (Table 11). For example, the commercial antimycotic (i.e., antifungal) agent Imazalil has been demonstrated to be effective when incorporated into low density polyethylene (LDPE) film for wrapping fruits and vegetables (Ben-Yehoshua et al., 1987). The same compound is effective at preventing mold growth on cheese surfaces when incorporated into LDPE films in contact. Although Imazalil is not approved for cheese packaging in the United States, the research results established that antimycotic films could be effective for control of surface mold on foods. The antifungal agent benomyl, commonly used as a fungicide, may be chemically coupled to ionomer film to inhibit microbial growth. It is improbable that benomyl would be approved for food use because of the chemical's toxicology.

Reports have appeared that demonstrate the effectiveness of adding common food-grade antimycotic agents to cellulose-based edible films. For example, films were constructed of cellulose derivatives and fatty acids to control the release of sorbic acid and potassium sorbate. Such films would seem to have the greatest application as fruit and vegetable coatings. Cellulose films, e.g., cellophane, generally are not heat-sealable and are not good water-vapor barriers in situations of high relative humidity. Cellulose films are generally obsolete today.

In general, several factors, including those dealing with human safety, should be considered in developing antimicrobial films:

- What is the spectrum of microorganisms against which the package might be effective? Films that may inhibit spoilage without affecting the growth

TABLE 11. Some Antimicrobial Agents of Potential Use in Food Packaging.

Class	Examples
Organic acids	Propionic, benzoic, sorbic
Bacteriocins	Nisin
Spice extracts	Thymol, *p*-cymene
Thiosulfinates	Allicin
Enzymes	Peroxidase, lysozyme
Proteins	Conalbumin
Isothiocyanates	Allylisothiocyanate
Antibiotics	Imazalil
Fungicides	Benomyl
Chelating agents	EDTA
Metals	Silver
Parabens	Heptylparaben

Source: Hotchkiss, 1995

of pathogenic microorganisms will raise safety questions analogous to those of technologies such as modified atmosphere packaging (MAP).

- What are the effects of the antimicrobial additives on the mechanical and physical properties of the plastic package material or structure?
- Is the antimicrobial activity a reduction in growth rate (while still permitting eventual increases in cell numbers) or does it cause cell death (decline in cell numbers)?
- To what extent does the antimicrobial agent migrate to the food, and what, if any, are toxicological and regulatory concerns?
- What is the effect of food product composition? Some antimicrobial agents, for example, are effective only at acid pH, whereas others might require certain product characteristics or compositions (e.g., a_w, protein, and glucose) to be effective.

The technology of controlling undesirable microorganisms by incorporating or coating antimicrobial substances onto household products such as cloth, kitchen utensils and sanitary goods, toys, and industrial structures like water-treatment filters has attracted much attention in Japan and now in the United States The antimicrobial agents that are applied to those items require consideration of toxicological safety, even if the materials containing them do not come in direct contact with food, drugs, cosmetics, etc. For food-packaging applications or direct human contact, safety must be ensured. The U.S. Environmental Protection Agency (EPA) recently has been questioning the efficacy and safety of such compounds in toys, kitchen utensils, and other items (Table 12).

Generally, metallic ions of silver and copper, the quaternary ammonium salts, and natural compounds such as Hinokitiol are regarded in Japan (but not in the United States) as safe antimicrobial agents. In food-use materials, the agents should not only be safe but also migrate with difficulty into foods. Silver-substituted zeolite (Ag-zeolite) and Microban® are probably most commonly used as antimicrobial agents incorporated into plastic materials.

ANHYDRIDES AS ANTIMICROBIAL AGENTS

Propionic acid, a commonly accepted food antimycotic agent, was coupled by Hotchkiss and his group to polyolefinic films, but antimycotic activity could not be demonstrated on rigorous testing. Direct addition of simple antimycotic agents such as propionic, benzoic, and sorbic acids to polymers such as LDPE was not successful because of lack of compatibility between the acid and the nonpolar film, probably due to differences in polarity. This problem was resolved by first forming the anhydride of the acid, which removes the ionized acid function and decreases polarity. Anhydrides are stable when dry and relatively thermally stable, yet become hydrolyzed in aqueous environments

TABLE 12 Applications of Antimicrobial Agents on Non-Packaging Solid Materials.

Textile and cloth	• Bed sheets
	• Towels
	• Socks
	• Shoes
Kitchen utensils	• Chopping boards
	• Baskets
	• Water purifiers
	• Scrubbing brushes
	• Trash buckets and liners
Sanitary goods	• Toothbrushes
	• Humidifier filters
	• Masks
	• Dust cloths
Food package materials	• Trays
	• Pouches
	• Paper containers
	• Drip absorbers
	• Overwrap films
Other applications	• Filters
	• Home electric appliances
	• False teeth
	• Mats for plant nurseries
	• Sandboxes for children
	• Toys

such as foods. Hydrolysis leads to formation of the free acid, which, in turn, leads to migration from the surface of the polymer to the food, where the free acids can function as effective antimycotics.

In research by Hotchkiss (1995), LDPE films into which benzoic anhydride was incorporated exhibited antimycotic activity when in contact with microbiological media and cheese. Benzoic anhydride, which had been added to LDPE film, was hydrolyzed within five hours and detected as benzoic acid in potato dextrose agar (PDA) and cheese after contact with the film. LDPE films, into which 1% benzoic anhydride was incorporated, completely inhibited *Rhizopus stolonifer, Penicillium* sp., and *Aspergillus toxicarius* growth. Lower concentrations of anhydride partially retarded growth by extending the lag phase and reducing the rate of growth in most cases. LDPE films containing added 0.5% to 2% benzoic anhydride delayed mold growth on cheese. These data suggest that addition of antimycotic agents to LDPE during film manufacture may be a feasible way of controlling surface mold growth in foods such as cheese. This is an example of "switched on" or truly active packaging: the active ingredient remains in the film until the film contacts the food; the activity is initiated by the moisture in the food.

BACTERIOCINS

A group of antimicrobial substances known as bacteriocins—proteins derived from microorganisms—are effective against microorganisms such as *Clostridium botulinum*. One such compound, nisin, has been accepted by regulatory authorities in some countries, such as Japan, for food use. These peptide-type compounds can theoretically be attached to the surface of food-contact films. It has not been reported whether such bound bacteriocins would be effective.

ANTIMICROBIAL ENZYMES

Antimicrobial enzymes might also be bound to the inner surface of food-contact films to produce microbial toxins. Several such enzymes exist, such as glucose oxidase, which forms hydrogen peroxide—a potent antimicrobial.

RADIATION-EMITTING FILM

A third possibility for antimicrobial films is to incorporate radiation-emitting materials into food-packaging films. Japanese researchers have reportedly developed materials that emit long-wavelength infrared (IR) radiation on exposure to water or water vapor. This finding is reported in Japanese publications to be effective against microorganisms without the risks associated with higher energy radiation. However, little direct evidence for the efficacy of this technology has been published in the English-language scientific literature.

ANTIMICROBIAL ACTIVITY OF SILVER IONS

The antimicrobial agents with the greatest potential appear to be those containing releasable silver salts. Antimicrobial activity of metals is due to the minute quantity of ions formed from the metals.

Copper ions can destroy microorganisms and viruses, and copper is indispensable for life as a constituent of metallic enzymes. Copper is not concentrated by animals and thus has few adverse effects on higher animals, which renders the ion relatively safe among metals. Recently, water filters and daily utensils utilizing antimicrobial activity of copper ion have been commercially available. Copper is regarded as toxic in contact with food, however, and generally is no longer permitted as an additive by regulatory authorities. Further, copper is an oxidation catalyst and thus would accelerate biochemical deterioration within foods.

Among metallic ions, the silver ion has the strongest antimicrobial activity (Table 13). Metallic silver does not release the ion easily, compared with copper, and so its antimicrobial activity is not quite as strong in its metallic state. Silver is a safe and relatively inert metal and therefore often used in direct human contact as dishes, plates, forks, spoons, knives, and tooth fillings.

Silver is used as an antimicrobial agent in the form of medicine and water-treatment agent. Silver ions can be dissolved in water, but they easily become insoluble by reacting with halogens, which is why silver's distribution in nature is limited. Silver is strongly absorbed by magnesium oxide, clay materials, and organic metal compounds. No reports exist with regard to carcinogenicity and mutagenicity of silver. In the United States, the standard for silver content in drinking water has been set at less than 50 ppb on the basis of a silver-containing medicine that causes angina symptoms.

ANTIMICROBIAL ACTIVITY OF SILVER NITRATE

Silver nitrate that forms silver ions in water solution has strong antimicrobial activity. Activity at much lower concentration than the antimicrobial level causes protein denaturation. Consequently, silver nitrate has a history of use as a therapeutic for bacterial infection and as an antiseptic in hospital environments. Silver is considered to interfere with metabolic functions of respiratory and electron-transport systems of microorganisms and mass transfer across cell membranes.

TABLE 13 Minimum Inhibitory Concentration
of Ions to *Salmonella typhi* at 37°C.

Ion	MIC
Na^+	1 0%
K^+	1 0
NH_4^+	1 0
Li^+	0 5
Sr^{++}	0 5
Ca^{++}	0 5
Mg^{++}	0 25
Ba^{++}	0 25
Mn^{++}	0 12
Zn^{++}	0 001
Al^{+++}	0 001
Fe^{++}	0 001
Pb^{++}	5.0×10^{-4}
Ni^{++}	$1 2 \times 10^{-4}$
Co^{++}	$1 2 \times 10^{-4}$
Au^{++}	$1 2 \times 10^{-4}$
Cd^{++}	$6 0 \times 10^{-5}$
Cu^{++}	$1 5 \times 10^{-5}$
Ag^+	2×10^{-6}

The antimicrobial activity of silver ions has been studied in relation to bacterial leaching and mining where silver ion inhibits growth of bacteria useful in leaching. Growth of the sulfur bacterium *Thiobacillus ferrooxidans* can be, slightly inhibited by silver nitrate at a concentration of 0.1 ppm, and the growth is completely suppressed at 1.0 ppm. The silver ion is first adsorbed by the surface of microbial cells and incorporated within the cells by active transport, inhibiting a range of metabolic enzymes, to demonstrate antimicrobial activity. Since silver ions react with proteins, they may react with different enzyme proteins after incorporation in the microbial cells and thus inhibit metabolic processes necessary for sustaining life.

Results reported from experiments with yeast inhibition indicate that Ag-zeolite's antimicrobial activity (see the next section) is observed under both aerobic and anaerobic conditions. The degree of activities on yeast is almost the same regardless of the oxygen concentration and the presence of light Thus, active oxygen does not seem to be directly involved.

Although silver demonstrates a fairly broad spectrum of antimicrobial activity, some bacteria resistant to silver and that absorb silver into the cell have been discovered.

ZEOLITES

In its crystalline structure, the mineral zeolite contains sodium ions that can be substituted by other metallic ions. For example, the Ag^+ ion can effectively replace the Na^+ ion to form Ag-zeolite. In the manufacture of Ag-zeolite, synthetic zeolite is normally used.

The lower the concentration of nutrient in media, the lower the concentration of Ag^+ ion required to demonstrate antimicrobial activity Activity is noticable even at concentrations of 0.02 to 0.05 ppm of Ag^+ ion (Table 14). The

TABLE 14. Effects of Concentrations of Ag-Zeolite and Dilution Times of Culture Medium on the Growth of *Saccharomyces cerevisiae*.

Ag-Zeolite (ppm)	Count at Dilution			
	1000	100	10	1
0	28×10^3	10×10^6	21×10^7	
1	29×10^3			17×10^8
5	40×10^2	12×10^4		
10	No growth	30×10^3	14×10^7	
100		No growth	No growth	98×10^7
1,000			No growth	No growth
5,000				No growth

Source: Anonymous, 1991a

release of Ag^+ ions from Ag-zeolite powder is not observed in pure water, but in nutrient media almost all the Ag^+ ions are released. It is suggested that the released Ag^+ ions react with sulfur compounds or other active constituents in media, and only a part of the released Ag^+ ions demonstrate antimicrobial activity. When the antimicrobial activity of Ag-zeolite containing 2.55 ppm of Ag^+ is determined in nutrient media, Ag^+ at the concentration of 1.5 to 3.5 ppm suppresses microbial growth (Table 15). A higher degree of substitution of Ag^+ ions in zeolite increases the activity against microbes.

Zeolite retains Ag^+ ions in stable and effective condition to facilitate their antimicrobial activity against the microorganisms with which they may come in contact. The unique feature of Ag-zeolite's activity is the broad antimicrobial spectrum of bacteria, i.e., little specificity of bacterial genera. Ag-zeolite is effective almost equally against bacteria, yeast, and mycelium fungi. This fact suggests that the action of Ag^+ ions is the inhibitory action on the components or functions that many different microorganisms share. Ag-zeolite does not demonstrate any activity against spores of heat-resistant bacteria, but rather only against vegetative cells. Thus Ag-zeolite has an important drawback in situations in which this class of microorganisms represents a hazard.

Ag-zeolite can maintain its antimicrobial activity only when it retains Ag^+ ions in the skeleton of zeolite; i.e., zeolites alone are not effective antimicrobials; rather they function as carriers. This activity is lost if all Ag^+ ions are eluted or if Ag^+ ions are inactivated by reacting with media constituents to form inert substances. In various foods many substances such as sulfates, hydrogen sulfide, and sulfur-containing amino acids react and weaken the activity of Ag^+ ions and so are capable of weakening Ag-zeolite's activity at ambient temperatures.

THE APPLICATION OF SILVER-ZEOLITE TO PACKAGE MATERIALS

Different kinds of antimicrobial package materials in which Ag-zeolite is incorporated into the plastics have been developed and tested for applications. Because Ag-zeolite is expensive, it is usually laminated as a thin, 3- to 6-μm (0.0002- to 0.0003-inch) coextruded layer containing Ag-zeolite, or the film can be applied on the surface of formed containers. The antimicrobial activity is demonstrated by the Ag^+ ions contained in zeolite particles on the film surface (Figure 1). Therefore, thicker film may not have an effect on the activity of Ag^+ ions inside the film at all. The amount of Ag-zeolite added may influence the heat-seal strength and other physical properties of the film, such as transparency. The normal incorporation level is 1% to 3%, but up to 5% has been tested.

All bacteria were eliminated in 1 to 2 days on the 1% to 3% Ag-zeolite-containing films, whereas the control film retained the initial viable counts in most cases (Tables 16, 17, 18, and 19). Bacteria may disappear quite rapidly

TABLE 15. **Growth-Inhibitory Effects of High-Exchange Ag-Zeolite Concentrations on *Saccharomyces cerevisiae* under Aerobic and Anaerobic Conditions.**

	Incubation	25 ppm	20 ppm	15 ppm	10 ppm	5 ppm	0 ppm
					Count at Concentration		
Ag-40	Aerobic	1.5×10^0	3.8×10^0	5.3×10^2	7.1×10^2	2.4×10^4	2.4×10^2
	Anaerobic	1.5×10^0	1.5×10^0	4.1×10^1	6.4×10^2	1.4×10^3	1.4×10^1
Ag-70	Aerobic	0	0	8.2×10^1	1.4×10^2	6.0×10^3	1.9×10^2
	Anaerobic	0	0	1.0×10^0	3.7×10^2	1.1×10^3	1.4×10^3
Aj-40	Aerobic	0	0	3.4×10^2	6.7×10^2	3.0×10^3	1.9×10^2
	Anaerobic	0	0	1.0×10^0	3.6×10^2	7.4×10^2	1.4×10^3
Aj-70	Aerobic	0	0	0	0	6.0×10^0	1.9×10^2
	Anaerobic	0	0	0	0	8.6×10^2	1.4×10^3

Ag = silver-exchanged; Aj = silver and zinc-exchanged.
Source: Anonymous, 1991a.

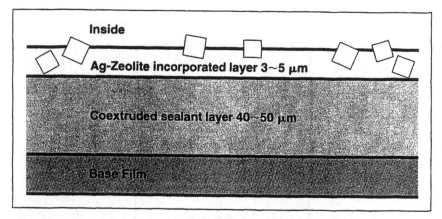

Figure 1 Structure of laminated film with Ag-zeolite Source: Vermeiren, 2000

when they are closely attached to the films or when the films do not contain any nutrients.

When microorganisms come into contact with the inner surface of the plastic, they take up the silver of silver-bonded zeolite, which disturbs their metabolic pathways and kills them. The presence of 1% silver zeolite in polyethylene is sufficient to reduce the microbial count on the surface of the plastic from 105 to 106 cells/ml to less than 10 cells/ml after 24 hours.

TABLE 16 The Antimicrobial Effect of Plastic Film Containing Silver Zeolite.

Bacteria	Film[1]	Microbiological Count After Time	
		0 Hours	24 hours
Escherichia coli	1	$7\,5 \times 10^5$	$1\,3 \times 10^6$
	2	$7\,5 \times 10^5$	<10
	ref	$7\,5 \times 10^5$	$1\,5 \times 10^5$
Staphylococcus aureus	1	$5\,8 \times 10^5$	$6\,7 \times 10^5$
	2	$5\,8 \times 10^5$	<10
	ref	$5\,8 \times 10^5$	$1\,5 \times 10^5$
Salmonella gallinarum	1	$5\,8 \times 10^5$	$1\,5 \times 10^5$
	2	$3\,6 \times 10^5$	<10
	ref	$3\,6 \times 10^5$	$4\,6 \times 10^6$
Vibrio parahaemolyticus	1	$1\,8 \times 10^5$	$7\,0 \times 10^5$
	2	$1\,8 \times 10^5$	<10
	ref	$1\,8 \times 10^5$	$5\,8 \times 10^5$

[1] 1:PE(25 μ)/PE (5 μ); 2:PE(25 μ)/PE(5 μ) with 1% silver zeolite; ref : PE(30 μ) (30 μ = 0 0012 inch, 25 μ = 0 001 inch, 5 μ = 0 0002 inch)
Source: Anonymous, 1991a

TABLE 17   Antimicrobial Effects of Ag-Zeolite Incorporated in Polyethylene Film.

Microorganisms	Sample	Microbiological Count After Time		
		0 Hours	24 Hours	48 Hours
Escherichia coli	a	$1\,7 \times 10^5$	<10	<10
	b	$1\,5 \times 10^5$	$5\,0 \times 10^6$	$4\,0 \times 10^5$
Staphylococcus aureus	a	$1\,0 \times 10^5$	$2\,6 \times 10^6$	<10
	b	$1\,1 \times 10^5$	$4\,6 \times 10^3$	$8\,7 \times 10^4$
Salmonella typhimurium	a	$2\,8 \times 10^4$	$3\,2 \times 10^2$	<10
	b	$3\,6 \times 10^4$	$3\,6 \times 10^6$	$4\,4 \times 10^6$
Vibrio parahaemolyticus	a	$2\,8 \times 10^4$	<10	<10
	b	$1\,7 \times 10^4$	$1\,6 \times 10^3$	$5\,6 \times 10^4$

(a) Polyethylene film incorporated with 1% Ag-zeolite (b) polyethylene film without Ag-zeolite incorporation
Source: Anonymous, 1991a

INTERACTIONS BETWEEN ANTIMICROBIAL ACTIVITY AND FOODS

When a plastic film containing Ag-zeolite is in contact with culture media, there should be a balance between cell growth rate and death rate of cells in contact with Ag^+ ion. Consequently, film containing 1% to 3% of Ag-zeolite does not display any antimicrobial activity in nutrient-rich culture media. To enhance the antimicrobial activity, the use of plastic materials with high substitution of Ag-zeolite or the application of Ag-zeolite on the film surface to increase the contact surface area may be another useful approach.

The silver substituted in zeolite molecular matrix may be eluted by the amino acids in foods. This effect was found to be different depending on the amino acid involved. There are essentially three types of amino acid groups and related compounds on the basis of Ag^+ elution pattern and their influence on antimicrobial effects (Figure 2). The glycine-type amino acid stimulates the elution of Ag^+ from Ag-zeolite but does not interfere in the action of Ag^+ ions, because its reaction with Ag^+ is weak. With lysine, the elution of and interaction with Ag^+ ions is relatively strong and therefore inhibits antimicrobial activity. With cysteine-type amino acids, both the elution and association with Ag^+ are strong and extremely inhibitory on antimicrobial activity.

These data show that certain kinds of amino acids and proteins influence the antimicrobial activity of Ag-zeolite. Therefore, it is necessary to consider the quality and quantity of amino acids and proteins in foods in terms of antimicrobial activity when films containing Ag-zeolite are applied to food-quality preservation.

DuPont has introduced MicroFree™ brand antimicrobial powders designed to impart antimicrobial activity to plastic resin systems. These products are

TABLE 18. Antimicrobial Effects of Ag-Zeolite Incorporated in Polyethylene Film on the Viable Count of *Escherichia coli* and *Streptococcus pneumoniae* in Chinese Oolong Tea.

Temperature		Sample	Microbiological Count After Time				
			0 Hours	5 Hours	24 Hours	48 Hours	
25°C	Ag-zeolite	Total count	1.7×10^6	4.4×10^5	2.5×10^3	3.3×10^1	
		Escherichia coli	3.6×10^5	1.2×10^4	<10	<10	
		Streptococcus pneumoniae	5.4×10^5	1.8×10^3	9.3×10^2	2.1×10^2	
	Control	Total count	1.7×10^6	6.5×10^5	7.6×10^4	3.9×10^7	
		Escherichia coli	3.6×10^5	1.9×10^4	5.0×10^3	1.0×10^6	
		Streptococcus pneumoniae	5.4×10^5	2.2×10^3	7.6×10^2	9.4×10^1	
10°C	Ag-zeolite	Total count	1.7×10^6	3.6×10^5	8.4×10^4	4.0×10^4	
		Escherichia coli	3.6×10^5	5.6×10^3	5.2×10^1	<10	
		Streptococcus pneumoniae	5.4×10^5	5.4×10^3	1.2×10^4	2.8×10^3	
	Control	Total count	1.7×10^6	7.1×10^5	1.5×10^5	2.3×10^4	
		Escherichia coli	3.6×10^5	5.8×10^3	8.0×10^2	1.6×10^2	
		Streptococcus pneumoniae	5.4×10^5	6.5×10^3	3.2×10^4	2.2×10^4	

Source: Anonymous, 1991a.

142

TABLE 19. Antimicrobial Effects of Ag-Zeolite Incorporated in Polyethylene Film on the Viable Count of *Pseudomonas aeruginosa* in Chinese Oolong Tea.

Temper- ature	Sample	Microbiological Count After Time			
		0 Hours	5 Hours	24 Hours	48 Hours
25°C	Ag-zeolite	92×10^5	51×10^5	<10	<10
	Control	92×10^5	55×10^5	23×10^4	<10
10°C	Ag-zeolite	92×10^5	79×10^5	25×10^5	26×10^4
	Control	92×10^5	81×10^5	26×10^5	86×10^4

Source: Anonymous, 1991a

claimed to suppress the growth of bacteria and mold responsible for unpleasant odors and discoloration in packaged food. Each of the three antimicrobial products has different attributes that allow it to be used in a range of polymers and coating systems. These photostable, inorganic powders, approved by the U.S. EPA (but not necessarily by the FDA), are reported by DuPont to function as antimicrobials in plastics. MicroFree™ utilizes a unique proprietary technology in which the overall loading of antimicrobial agents is reduced, providing environmental benefits and cost effectiveness. The reduced loading also reduces the impact on film color and opacity. These inorganic powders are reportedly effective in packaging. They are unaffected by high temperature, allowing them to be processed into a number of plastic resin systems. DuPont reports no volatility or odor problems during processing or in end-use applications.

The carrier particle serves to increase the surface area of the active ingredients so that the overall loading of antimicrobial agents is reduced. The barrier

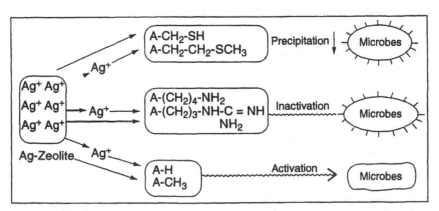

Figure 2 Effects of amino acid on the antimicrobial activity of Ag-zeolite Source: Anonymous, 1991a.

coating controls the rate of release of the active ingredients, providing the proper level of antimicrobial activity, reducing interaction with polymers, and enhancing photostability. A proprietary dispersion coating is also provided to facilitate dispersion of the powder in polymers.

Comparing its Microfree™ silver salt and other antimicrobial materials, DuPont indicates the following attributes:

(1) Antimicrobial coatings, in general:

- silver ions—antibacterial
- copper oxide—antifungal
- zinc silicate—antifungal

(2) Support vehicles or core particles:

- titanium dioxide—opacity
- barium sulfate—translucent
- zinc oxide—antifungal activity

Benefits reported by DuPont for the Microfree™ product are:

- broad-spectrum antimicrobial activity (bacteria, yeast, mold)
- low color generation
- low toxicity
- environmentally "friendly"
- low interaction with polymer matrices
- easy dispersion
- applicability for both food and cosmetic package materials

MicroFree™ powders have been demonstrated to be effective against the following microorganisms·

(1) Bacteria (gram-positive and gram-negative)

- *Escherichia coli*
- *Pseudomonas aeruginosa*
- *Klebsiella pneumoniae*
- *Staphylococcus aureus*

(2) Fungi

- mold—*Aspergillis niger*
- yeast—*Candida albicans*

Currently available MicroFree™ powder grades include:

- Z-200—silver on a zinc oxide core
- T-558—silver, copper oxide, and zinc silicate on a titanium dioxide core
- B-558—silver, copper oxide, and zinc silicate on a barium sulfate core

MicroFree™ is claimed to be engineered for durable antimicrobial activity. It is an inorganic particle unaffected by solvents and temperatures up to 325°C. MicroFree™ antimicrobial activity occurs as the microbe is exposed to low levels of silver and zinc oxide, requiring 1 to 24 hours, depending on the application and concentration. MicroFree™ imparts activity at the surface of a plastic. Because of its temperature stability, MicroFree™ can be master-batched to 40% for later incorporation in the final system. With its proprietary dispersion aid and barrier coating, MicroFree™ is dispersed in polymers systems with little or no impact on the physical properties.

EXAMPLES OF ZEOLITE PRODUCTS

In Japan, antibacterial agents, and particularly zeolites with metal ions, are incorporated into package materials. Zeolites are minerals of the form $xMn_2O \cdot Al_2O_3 \cdot ySiO_2 \cdot zH_2O$. Silver, copper, or zinc is added to natural or synthetic zeolites. These minerals exchange with metal ions (i.e., sodium), preferably in combinations such as AgCu and AgZn, for improved resistance to temperature for plastic film applications. These are multiporous structures with 3 to 8 Å diameter openings. With particle sizes of 0.5 to 1.0 micron, they have no effect on film transparency and heat sealability at the 1% to 3% level at which they are used. As antibacterial films, zeolites containing salts are reported to be particularly effective in protecting fish and meats against fungi and other microorganisms.

In Japan, many antibacterial package types currently are on the market. The following are some examples:

(1) Adding zeolite to package materials is claimed to keep contents fresher by entrapping microorganisms and absorbing gases such as ethylene. Japan's Tadashi Ogawa has developed a low density polyethylene sheet that they claim can preserve food that is wrapped in it. Ogawa asserts that the additive absorbs infrared radiation and re-emits it at a specific frequency destructive to bacteria.

(2) Japan's Nippon Unicor markets polyethylene packaging film incorporating a ceramic additive.

(3) The primary microbicidal agent of Japan's Sangi's Apacider-A® is stabilized silver. The stability is achieved by bonding silver chemically and physically with calcium phosphate on a zeolite support skeleton. Apacider-A is stable to light and heat and is processed to prevent excess release of silver during processing. Thus, Apacider products can overcome the disadvantages of other antimicrobial plastic additives by maximizing antimicrobial benefits while minimizing environmental impact (Table 20). Apaciders withstand very high temperatures, making them suitable for use

TABLE 20 Test Results on Apacider AW against *Escherichia Coli* in Polypropylene and Polyethylene.

| | | Microbiological Count After Time | | |
	Test Material	Start	24 Hours	48 Hours
In polypropylene	Control polypropylene	$1\,0 \times 10^5$	$3\,5 \times 10^7$	$4\,6 \times 10^7$
	0 5% Apacider AW in polypropylene	$1\,0 \times 10^5$	0	0
	Bacteria solution	$1\,0 \times 10^5$	$3\,5 \times 10^7$	$4\,6 \times 10^7$
In polyethylene	Control polyethylene	$1\,2 \times 10^5$	$2\,0 \times 10^5$	$5\,0 \times 10^5$
	0 5% Apacider AW in polyethylene	$1\,2 \times 10^5$	0	0
	Bacteria solution	$1\,2 \times 10^5$	$2\,0 \times 10^5$	$5\,0 \times 10^5$

with plastics. Apacider products are offered as very fine white powders that remain white even after incorporation into most commonly used plastic resins. Claims made by the maker for Apacider products are that they are effective additives with broad-spectrum antimicrobial benefits, and they display minimal toxicity to humans or animals.

Apaciders are designed to provide virtually zero silver ion (<0.5 ppm) dissolution and therefore demonstrate little or no halo or zone effect, as do organic or inorganic silver ion eluting agents. The proper and complete dispersion of Apaciders into the plastic resin is critical to achieve maximum particle density at the surface. Various surfactants may enhance dispersion and migration toward the surface.

In 1983, an antimicrobial zeolite, Zeomic®, was developed by Japan's Shinanen New Ceramics Company with Mitsubishi Corporation. As indicated earlier, zeolite is mainly composed of aluminosilicate matrix with a three-dimensional skeletal structure filled with various cations. Those cations could successfully be exchanged with antimicrobial silver ions as described above. Zeomic® is available in the form of fine white powder suitable for a wide range of commercial and industrial applications. Zeomic® is characterized by its nontoxicity, wide antimicrobial spectrum, high heat resistance, and long time of efficacy.

The antimicrobial effectiveness of Zeomic® was confirmed in Japan against more than 30 microorganisms including bacteria, molds, and yeasts. A resistance-acquiring capacity test was also conducted. In this test, *Staphylococcus aureus*, methicillin-resistant *Staphylococcus aureus*, and *Pseudomonas aeruginosa* were cultured in multiple generations and the results indicated almost no variation in the minimum inhibitory concentration (Tables 21, 22, and 23).

Since the antimicrobial metal ions are exchanged (bonded) ionically in Zeomic, only slight release of these ions is expected. Release tests performed

TABLE 21. Minimum Inhibitory Concentration (MIC) for Zeomic® (Unit: ppm).

Test Specimen	MIC (ppm)	Test Specimen Characteristics
Bacillus cereus	25	Spore-forming bacteria in milk and cream products
Escherichia coli	62 5	Bacterial strain of the intestinal flora of mammals, which can cause urinary tract infections, food poisoning, etc
Pseudomonas aeruginosa	62 5	Bacteria widely distributed in nature and the hospital environments, and considered as an opportunistic pathogen
Staphylococcus aureus	250	Bacteria causing wound infections, food poisoning, abscesses, and toxic shock
Streptococcus faecalis	125	Occasionally opportunistic pathogenic bacteria
Vibrio parahaemolyticus	62 5	Bacteria widespread in inshore marine waters; occurring naturally on seafood other than shellfish
Aspergillus niger	500	Mold used in the industrial production of citric and gluconic acids
Penillium funiculosum	500	Mold strain causative of spoilage of vegetables, bread, etc
Candida albicans	250	Usually nonpathogenic yeast, which can invade susceptible host, causing candidiasis
Saccharomyces cerevisiae	250	Yeast widely used as baker's yeast and the commercial production of wine and ethyl alcohol

Source: Anonymous, 1991b

with water reveal that only small quantities (order of magnitude of parts per billion) migrated to the suspending medium. This feature implies that the presence of the antimicrobial ions could be expected for long periods. In addition to this, the fact that Zeomic neither vaporizes nor decomposes supports long antimicrobial activity.

Generally, organic antimicrobial agents do not have high heat resistance, and so their uses are limited. In contrast, the inorganic nature of Zeomic permits better thermal stability even at temperatures up to 800°C, thus suggesting a wide range of applicability.

Silver ions may be loaded onto Chabozite, a naturally occurring zeolite This antimicrobial mineral may then be incorporated into the extrusion of film or coated on the inside of film so that contact of the film interior with the product contents could destroy microorganisms.

TABLE 22. Antimicrobial Effect of Zeomic® Incorporated in Plastic Resin (Acrylonitrile Butadiene Styrene, Polypropylene, Low-Density Polyethylene).

Bacterial Strain	Specimen	Microbiological Count After Time	
		0 Hour	24 Hours
Escherichia coli	1	$2\,6 \times 10^4$	Less than 20
	2	$2\,4 \times 10^4$	$3\,0 \times 10^4$
	3	$2\,3 \times 10^4$	Less than 20
	4	$2\,3 \times 10^4$	$2\,4 \times 10^4$
	5	$2\,5 \times 10^4$	Less than 20
	6	$2\,4 \times 10^4$	$2\,8 \times 10^4$
	Control	$2\,4 \times 10^4$	$3\,2 \times 10^4$
Staphylococcus aureus	1	$5\,1 \times 10^4$	Less than 20
	2	$5\,0 \times 10^4$	$4\,7 \times 10^4$
	3	$5\,1 \times 10^4$	Less than 20
	4	$5\,1 \times 10^4$	$5\,3 \times 10^4$
	5	$5\,2 \times 10^4$	Less than 20
	6	$5\,2 \times 10^4$	$5\,1 \times 10^4$
	Control	$5\,1 \times 10^4$	$5\,6 \times 10^4$
Pseudomonas aeruginosa	1	$1\,6 \times 10^4$	Less than 20
	2	$1\,8 \times 10^4$	$2\,1 \times 10^4$
	3	$1\,8 \times 10^4$	Less than 20
	4	$1\,9 \times 10^4$	$2\,4 \times 10^4$
	Control	$1\,8 \times 10^4$	$2\,3 \times 10^4$

Specimen 1: Acrylonitrile butadiene styrene (ABS) with Zeomic® 1%; 2: ABS without Zeomic 1%; 3: Polypropylene (PP) with 1% Zeomic; 4 PP without Zeomic; 5: Low-density polyethylene (LDPE) with 1% Zeomic; 6 LDPE without Zeomic
Source: Anonymous, 1991b

TABLE 23. Antimicrobial Effect of Zeomic® Incorporated in Polyethylene (PE) Film.

Bacterial Strain	Specimen	Microbiological Count After Time		
		0 Hours	24 Hours	48 Hours
Escherichia coli	1	$1\,7 \times 10^5$	Less than 10	Less than 10
	2	$1\,5 \times 10^5$	$5\,0 \times 10^6$	$4\,0 \times 10^6$
Staphylococcus aureus	1	$1\,0 \times 10^5$	$2\,6 \times 10^2$	Less than 10
	2	$1\,1 \times 10^5$	$4\,6 \times 10^4$	$8\,7 \times 10^4$
Salmonella gallinarum	1	$2\,8 \times 10^4$	$3\,2 \times 10^2$	Less than 10
	2	$3\,6 \times 10^4$	$3\,6 \times 10^6$	$4\,4 \times 10^6$
Vibrio parahaemolyticus	1	$2\,8 \times 10^4$	Less than 10	Less than 10
	2	$1\,7 \times 10^4$	$1\,6 \times 10^3$	$5\,6 \times 10^4$

Specimen 1: Polyethylene with Zeomic® 1%; 2: Polyethylene without Zeomic
Source: Anonymous, 1991b

148

Japan's Chugoku Pearl developed a thermoplastic resin (e.g., polypropylene), containing antimicrobial zeolite (0.1% to 5 wt.%), plus copper ions and laminated to paper. The resulting sheet is intended to wrap food.

Japan's Try Company and Taiyo Chemical Company jointly developed a pouch claimed to preserve drinking water for the long term. The pouch is constructed of transparent film composed of a five-layer coextrusion of nylon/polyethylene film. The inner layer of this coextruded film is antibiotic by virtue of incorporation of Ag^+ zeolite as inorganic antibiotic materials. Because of their fixed porous crystal structure, however, their diameters as fine powder are limited to those that decrease light transmission of film and sheet containing the fine powder. For this reason, traditional film pouches that contain zeolites cannot be optically inspected to determine the condition of the drinking water contents. According to the suppliers, when water is packaged in polyethylene terephthalate polyester bottles, which are, of course, transparent, the long-term safety of the contents cannot be guaranteed (Anonymous, 1997).

This new pouch incorporates a type of silver ion inorganic antibiotic material smaller than conventional zeolites, allowing it to be contained in the inner polyethylene film layer. The content of the antibiotic silver in the film layer is 1% to 5% by weight. When water containing 14,000 cfu/ml microorganisms is filled into the new antibiotic pouch, 100% of the microorganisms are claimed to be destroyed within 24 hours.

This pouch has good transparency because of the diameter of the particles; thus, water content conditions can be optically inspected. After opening, the antibiotic layer is claimed to prevent microbiological propagation.

The venture anticipated selling the pouch under the tradename "Miracle Water Pack." Miracle Water Pack is designed so that the water content continuously undergoes natural convection movement because of temperature changes of the external atmosphere. Water, undergoing natural convection and in contact with the inner antibiotic mineral layer, is claimed to produce hydroxyl radicals that destroy microorganisms.

Bacte Killer, a product of Japan's Kanebo, is an aluminum silicate-based synthetic zeolite containing both silver and copper ions. Two types are offered: one in which the Na^+ ions have been replaced with both Ag^+ and Cu^+, and the other in which only silver is used. This material has been reported to be safe for food packaging by the Japanese Food Analysis Center. One type, containing particles 2 μm in diameter, is intended for incorporation in film and containers for food packaging.

SILVER ION ANTIMICROBIALS

Novaron, offered by Japan's Toa Gosei Chemical Company, is a high heat-resistant inorganic silver material for the fibers and plastics industries. It is currently incorporated into polyester, acrylic, and nylon fibers and into plastic molding compounds for kitchen wares and bathroom products. Reportedly it has a deodorizing effect.

Nimiko Company of Osaka, Japan, markets Silvi film, which employs silver oxide ions (AgO_2^+) for anti-mold and antibacterial applications. It is a combination of silver and SiO_2, blended with plastic film that allows for the gradual release of the AgO_2^+ ions. Fresh green vegetables, which emit water vapor when they respire, are an example of a product that might be protected. It is not necessary for the packaged food to directly contact the Silvi film, as the latter is claimed to be effective even if the solid food is immersed in a liquid or otherwise not in direct contact with the film. The material has been judged safe for food use by the Nippon Food Analysis Center, not an official government agency. The material is also heat and light resistant, and its activity is claimed to not decrease with time. Nimiko blends silver ions with single-layered sheets of polyethylene in standardized thicknesses, coextruded multilayer films, and adhesive-laminated films. Various pouch types will accept an inner layer of Silvi film. The product has been found to be especially effective with fresh meat in vacuum packaging.

Japan's Daikoku Kasei has a similar product to that in Silvi Film, called Piatech. Piatech is effective in high relative-humidity environments in preventing mold and bacteria generation. It reportedly meets Japanese Food Sanitary Law requirements.

ADDITIONAL MINERALS

In Denmark, work has been performed with diatomaceous earth incorporated into film and used as a poultry-tray soaker pad. Unidentified antimicrobial agents were incorporated into the diatomaceous earth to destroy microorganisms (Hansen et al., 1989).

ADDITIONAL ANTIMICROBIALS

A host of other antimicrobials besides zeolites has been proposed and evaluated for incorporation into package materials, most of which do not have food-regulatory acceptance.

MICROBAN®

Microban® is a triclosan-based [5-chloro-2-(2,4-dichlorophenoxy) phenol] antimicrobial that has been widely employed in polyolefin applications such as solid waste bags, pails, cutting boards, and toys. In the United Kingdom, such biocides are used in polyolefin delicatessen wraps and reheatable food containers. Microban® technology was first developed for the medical arena and is now claimed to be in development for hotel/restaurant/institutional and food-processing-packaging applications.

The triclosan-based antimicrobial agents used in polymer applications (both flexible films and rigid plastics) are sub-micron-sized cell-wall penetrants that disrupt the metabolic function of microorganisms such as bacteria, yeasts, and fungi. When introduced into the molecular structure of polymers, these additives attack microorganisms, interrupting their ability to function, grow, and reproduce. They are claimed to deliver continuous control of a broad range of Gram-positive bacteria (e.g., *Staphylococcus aureus*) and Gram-negative bacteria (e.g., *Escherichia coli*).

The antimicrobial additives are introduced into the interstitial spaces of the polymer matrix by means of a proprietary process and are claimed not to affect the physical properties of the plastic. The empty spaces act as reservoirs for the additives, which bloom to the surface as needed.

Key to their functionality is the agents' ability to attach only to thin-walled cells Thick-walled cells, such as those in humans and animals, are claimed to be unaffected by the additives. The agents are claimed to be safe for contact with human skin Nevertheless, as indicated above, Microban® is not accepted by U S. regulatory authorities for food package material in contact with foods. Further, Microban® is being questioned by the U.S. Environmental Protection Agency (EPA), which has jurisdiction over consumer products when they are marketed with claims that they destroy microorganisms on inanimate objects. There appears to be overlapping responsibility for EPA and FDA in the area of microbicides. EPA is reportedly moving forward with their administrative oversight. FDA's position is not clear.

Microban® technology could be employed in cast, coextruded, injection-molded, or blow-molded plastic products. The company claims success in its proprietary technology in polyethylene and polypropylene blown and cast films, with the applications being surgical drapes and textiles (Rubinstein, 2000).

BENOMYL

Japan's Sumitomo offers thermoplastic resin food-packaging films that are rendered antibacterial and fungicidal by the incorporation of benzimidazole, phthalimide, and sulfamide active ingredients. A fungicide has been chemically attached to a food-packaging polymer chain so that the preservative retains its potency. "Benomyl" is one of the most widely used systemic fungicides; its breakdown product, methyl-2-benzimidazolecarbamate, is also fungicidal. Growth of *Aspergillus flavus* and *Penicillium notatum* molds on agar was inhibited by benomyl incorporated into a 0.001-inch gauge ionomer film. A chemical method was used to attach the fungicide (containing amino groups) to the ionomer (containing pendant carboxyl groups): dicyclohexyl-carbodiimide was used as the coupling agent and dimethylaminopyridine as the reaction catalyst. As indicated above, Benomyl will probably not be accepted by U.S. regulatory authorities for food contact.

SORBIC ACID

Sorbic acid has been used for years as an antimycotic, i.e., mycostatic, agent in foods. A process has been developed for the production of fungicidal paper for long-term storage of bread. Vegetable parchment was used as the base paper, whereas sorbic acid was used as the fungicide. The active compound was coated on the surface in the form of an aqueous suspension, using carboxymethylcellulose (CMC) as a binder to prevent the shedding of crystals from the paper surface. The effects of sorbic acid particle size (less than 20 μm) and the amount of CMC (5% to 15%) on the protective properties of the paper for preserving bread for six to eight months were determined. The results indicated a suppression of mold as a result of contact with the paper.

Purdue University researchers have studied mechanisms responsible for releasing antimicrobial agents from package materials by measuring diffusivity and other mass transfer properties of potassium sorbate through plastic films (e.g., low- and high-density polypropylene, polypropylene, biaxially oriented PET). Effects of temperature (4°, 25°, and 50°C) (39°, 77°, and 122°F) and concentrations of potassium sorbate (1%, 25%, and 40%) on mass transfer were also estimated. Diffusion of potassium sorbate through polymeric films was independent of concentration of potassium sorbate and followed the law of Fickian diffusion (Han and Floros, 1996).

Sorbic acid and its derivatives are accepted by U.S. regulatory authorities as in-package mycostatic agents.

Sorbic acid-treated crepe paper was found to be an effective fungistatic wrapper for bread. Incorporation of an antioxidant in the treated wrapper plus the use of an odor adsorbent such as charcoal inside the bread packages minimized off-flavor development. Sliced bread was found organoleptically acceptable up to one month, and as sandwiches, up to three months. Sorbic acid–treated wrappers could not be used for the preservation of cheese because of the production of bitterness or browning.

In other development work at Purdue University, the fungicide "Imazalil" was chemically coupled to plastic films to delay the growth of molds on contained foods. Another example was a zeolite film in which incorporated zeolite holds antibiotic compounds in a plastic matrix and slowly releases them.

NATURAL ANTIMICROBIALS

Viskase Corporation, (now part of Curwood, Division of Bennis Corporation) has reported on the application of natural antimicrobials to package materials to retard microbial growth on meat products Antimicrobial compounds naturally derived or occurring in nature have been isolated from plant and animal sources. The compounds from plant origin include extracts of spices: cinnamon, allspice, clove, thyme, rosemary, and oregano are a few that have shown antimicrobial activity Some plant extracts having similar effect on microbe growth are onion, garlic, radish, mustard, and horseradish. Other classes of naturally derived substances are produced from fungal and bacterial action. The most noted of those derived is the polypeptide nisin. Others include natamycin, pediocin, and various bacteriocins (Nicholson, 1997).

While some natural antimicrobials function by chelating or tying up key chemicals, most function to interrupt the metabolic pathway by interference in the cell wall membrane/structure. Nisin, for example, interacts with the sulfur-containing compounds in the bacterial membrane, disrupting their semipermeable function and causing lysing of the cells. Other cationic systems such as surfactants function similarly. Chemical systems like salt, glycol, and others perform due to a reduction of the water activity within the cell. Since water is a key compound in the metabolism of a cell, any disruption of the fluid flow causes growth suppression.

Antibacterials

Compounds frequently used as antibacterials include inorganic acids/salts, organic acid esters, and complex polypeptides. The nitrites and sulfites find use in a broad range of foods from meats to vegetables. Most common of the organic acid esters are the lactates, sorbates, and parabens. Of the polypeptides, nisin is the most common. Similarly produced inhibitors from bacterial cultures are called bacteriocins. Although they are not universally effective, bacteriocins are produced from a complex of different strains of lactic acid bacteria and impose hurdles to growth for the general microbiological population.

The target for an antibacterial system carried by a package material may be focused in two arenas: pathogens and the reduction of the normal spoilage bacteria such as *Lactobacillus* and *Pseudomonas* in meats.

Viskase's research originally was directed towards *Listeria monocytogenes* on processed turkey, ham, and beef. Their results indicate a two-log reduction over a 12-week period of storage. Active ingredients were nisin-like and pediocin-like compounds.

Work by Viskase with raw meats demonstrated that the naturally occurring antimicrobials were ineffective. Since bacteriocins are complex polypeptides, it is reasoned that the natural proteases in the uncooked meat destroy them before they can act on any bacterial contamination. Since most potential safety concerns are due to postprocessing contamination, any inadvertent contamination is on the food surface due to handling. By applying the natural antimicrobial formulation to the inside (food contact side) of the food package, the contaminated surface directly contacts the bacteriocin.

Synergistic Effects

To address the other target antimicrobials, Rhone-Poulenc developed an active flavor protectant ingredient to control spoilage bacteria called Micro-GARD™. An improvement over current single-hurdle biopreservation technology where either pathogens or spoilage organisms are being addressed would be a multiple-component system. By combining the best of both systems and applying them together through a packaging system, both pathogen-protection and storage life-improvement could result.

Just as chemical food additives, process parameters, and food formulation/ingredients represent additional hurdles to microbiological growth, a food package with an inherent mixture of biopreservatives (naturally occurring antibacterials) has a chance to suppress many species of bacteria.

Viskase research data have demonstrated that application of the natural antimicrobials produced from select lactic acid–producing bacteria can positively affect the shelf life and storage stability of many foods. By combining the synergistic effects of the natural antimicrobial nisin with a chelator, they demonstrated a two-log suppression of the pathogen *Listeria monocytogenes* in processed meat products over 12 weeks of storage.

Development continues to combine the multiple (i.e., synergistic) hurdle effect of more than one natural antimicrobial. It would appear that the natural antimicrobials function by extending the lag growth phase and may be potentiated by chelators and other interactive compounds.

VITAMIN E CONTROL OF RELEASE OF ANTIMICROBIAL

A 1996 patent (Seabrook and Heymann, 1996) described how vitamin E within plastic can regulate release of natural antimicrobial agents incorporated into plastic materials. Applications may include plastic sheets and film, molded containers, and plastic products. Applications anticipated by the patent include both food and non-food uses, depending on the antimicrobial agent used.

The patent states that ". . . chemical controllers are added to the compositions for controlling the rate of release of the antimicrobial agents. The chemical controllers include vitamin E. . . . Vitamin E controls the rate of migration of the antimicrobial agents so that harmful amounts of antimicrobial agents are not released."

CHLORINE DIOXIDE

Research at the Southwest Research Institute (San Antonio, Texas) sponsored by Bernàrd Technologies, Inc. (Chicago, Illinois) has resulted in a series of proprietary and patented biocidal polymeric films and coatings that generate sustained and controlled amounts of the biocide chlorine dioxide, thereby forming an active Microatmosphère™ system. The generation of chlorine dioxide within the package's interior from the solid-state technology is activated by one or more of the ambient environmental conditions, such as humidity. The technology shows particular promise for food packaging. The duration and amount of chlorine dioxide generated can be modulated to kill a broad range of microorganisms, including mold spores. These polymer films were successfully used to destroy *Escherichia coli* in ground fresh meat products (Table 24) (Wellinghoff, 1995).

In recent years, the FDA allowed several additional uses for chlorine dioxide (ClO_2) as a part of food-processing procedures. Chlorine dioxide has been recognized widely since the 1950s and 1960s as a potentially useful substance for the control of bacteria, viruses, and fungi. Current commercial applications, in addition to the food-released uses, include skin disinfection, care and cleaning of contact lenses, and the generation by municipalities of potable drinking water.

Chlorine dioxide is well known in the food-processing industry as a powerful broad-spectrum antimicrobial that is fast-acting, nonirritating, and nontoxic at the concentrations effective for allowed use. It does not form trihalomethanes (THMs) or dioxins. Because its mechanism of action is oxidative rather than chlorinating, it attacks multiple cellular sites, such as membranes and fundamental microbial cellular processes, and does not permit the selection of resistant microorganisms.

The controlled solid-state sustained generation system, developed by the Southwest Research Institute and Bernàrd Technologies, Inc., produces an initial level of chlorine dioxide at concentrations of 0.5 to 1 ppm or higher. This is followed by a longer period of sustained release when further biocidal activity is needed. The active components in the controlled-release design can be modulated to generate the chlorine dioxide quickly, in a matter of days or hours, or more slowly over a period of weeks or months.

TABLE 24. Effect of ClO_2 in Paper Separators Between Ground Beef Patties on *Escherichia coli* on Ground Beef.

Experiment	Initial Bacterial Load (cfu/Patty)	Thickness (Inches)	Density	Results Log (Reduction)	Results (% Reduction)
I	17,000,000	0 5	Loose	>6	>99 99%
II	50,000,000	0 75	Dense	2 to 3	99 5%
Infectious dose	<200			<1 cfu/patty	nd

Source: Wellinghoff, 1995

Temperature and atmospheric moisture affect the polymeric film, causing hydrolysis of an acid moiety that diffuses within the film to generate the chlorine dioxide in the form of a Microatmosphère™ system. As the conditions become more favorable for microbial growth, the rate of chlorine dioxide generation increases. Mold and bacterial growth is highly correlated to temperature and relative humidity.

Several different methods for polymeric film application have been developed. The system can be extruded directly into the polymer and made into film. The polymer also can either be sprayed or melt-coated onto a container material. An alternative approach is to adhere a "coupon" or label within a package to permit the chlorine dioxide to diffuse throughout the interior of the package containing such items as fruits or vegetables. The chlorine dioxide rapidly dissolves in the water on the surface of the package's contents.

Chlorine dioxide reportedly is effective at extremely low levels, on the order of a few ppm, and can kill molds, bacteria, and viruses at levels below which an odor can be detected (widely considered to be 10 ppm). The patented generation system can be optically clear, a result achieved through the development of a single-phase solution formed between the precursor chlorite and the polymer, resulting in particles small enough not to scatter visible light. This means that if clear plastic packaging is desired for a product, the polymer can be co-extruded with barrier films to make a clear, multilayered structure (Wellinghoff, 1995). Federal registration of chlorine dioxide-generating packaging components has been on an expedited fast track (Gray, 2000).

PHOTON-EXCITED NYLON FILM

At the University of Delaware, studies were conducted on an antimicrobial package that does not leach biocides into food. Nylon film was treated with ultraviolet irradiation to increase surface amine concentration (nylon is a polyamide), which acts as surface-active antimicrobial sites. The film was tested against *Staphylococcus aureus, Enterococcus faecalis,* and *Pseudomonas fluorescens* (Hagelstein et al., 1995). After 6 hours, a three-decimal reduction in *S. aureus* was found from an initial population of 8×10^3 cfu/ml. Sterility of the buffer was obtained when the initial population of *S. aureus* was 8×10^3 cfu/ml. The film was less effective against *E. faecalis* and *P. fluorescens.* The optimal temperature for inactivation was 37°C. A 90% reduction in counts was achieved in the stationary sample, and a 99% reduction was achieved in the agitated sample.

Ultraviolet (UV) photon irradiation from an excimer laser was used to impart the antimicrobial activity to the surface of polyamide films. A polyamide (nylon 6,6) film was treated with UV photons having a wavelength of 193 nm. The source of the photons was an excimer laser with photon energy imparted to the material at an irradiance of approximately 200 kW/cm² for 16 nanoseconds and a total dosage of 2 7 J/cm^2. Results from X-ray photoelectron spectroscopy and reactions of the irradiated surface with dyes used to titrate amine end groups in nylon, in the same way that it reacts with those of amines, indicate that the amide groups on the film surface are converted to amines.

The UV irradiation treatment increased surface amine concentration acting as surface-active antimicrobial sites. The film proved most effective against *S. aureus*, optimal temperatures for inactivation being 37°C.

ETHANOL

Ethanol is used routinely as a sterilizing agent in medical and pharmaceutical applications, and, in vapor form, it has been demonstrated to extend the shelf life of packaged bread and other baked products. Ethanol sachets, containing encapsulated ethanol, release ethanol vapor, which imparts a preservative effect in the packaging headspace. The ethanol prevents microbiological spoilage of intermediate moisture foods, cheese products, and sweet bakery goods (Smith, 1988).

Ethyl alcohol or ethanol has been known for a century to act as a microbicidal preservative and to be safe for human consumption when ingested in very small amounts (as in beverages). Difficulties exist, however, in applying ethanol as a preservative to foodstuffs: off-flavors, rapid volatilization, and consumer resistance.

Japan's Freund Industrial Co., Ltd., has explored ethanol's use in food preservation and developed a technology for microencapsulating food-grade ethanol. This patented process microencapsulates ethanol in a specially engineered silica that maintains a free-flowing powder form and contains a minimum of 55% ethanol by weight. Freund trademarked this material as Ethicap® or Antimold 102. Freund no longer offers its microencapsulated ethanol in the United States.

When food is packaged, moisture is absorbed from the food and ethanol vapor is released from its encapsulated form to permeate the package headspace and, to some degree, the product.

Ethicap® is supplied packaged in small sealed sachets composed of laminated material suitable for ethanol vapor release and conforming to FDA's food-packaging materials regulations. When an Ethicap® sachet is inserted in a food product package and sealed, Ethicap® microcapsules release ethanol vapor, thus filling the headspace of the food package and surrounding the food Food products packaged with Ethicap® can be preserved for periods of several weeks to several months, depending on the type of food and the amount of Ethicap® used (Table 25). In addition, as indicated earlier in the "aroma emissions" section, Freund claims food aromas can be added to Ethicap® to mask the ethanol aroma when the food package is initially opened or

TABLE 25. Ethanol Generators Used in Packaging.

Type	Function	Product Application	Water Activity of Products
Ethicap® (Antimold 102)	Generates ethanol vapors	Cakes and breads (moisture dependent)	>0 85
Negamold™	Generates ethanol vapors, absorbs oxygen	Cakes and breads (moisture dependent)	>0 85

Source: Smith, 1988

to enrich the flavors of the food products being preserved. This synergistic property of encapsulated ethanol may be applicable to enhance the attributes of foods.

Examples of foods currently being marketed with Ethicap® in Japan include:

- rice cake stuffed with bean jam
- steamed bun with bean jam
- cupcakes
- doughnut with bean jam
- soft rice cakes
- iced sponge cake

Ethicap® has been shown to be most effective in preserving intermediate-moisture foodstuffs with water activity levels between 0.7 and 0.9. The criteria in determining how to use Ethicap® effectively for shelf life extension are water activity and weight of the foodstuff, volume of the package, and the type of packaging laminate used.

Ethanol vapors have been shown to be effective in controlling molds including *Aspergillus* and *Penicillium* species; bacteria including *Staphylococcus, Salmonella,* and *E. coli* species; and three species of spoilage yeast. Ethanol vapors have been used extensively in Japan to extend the mold-free shelf life of high-ratio cakes and other high-moisture bakery products. A 5- to 20-times extension in mold-free shelf life has been observed for cakes. Results also have demonstrated that products containing sachets did not become as hard as the controls and, further, results were better than those using an oxygen scavenger to inhibit mold growth, indicating that the ethanol vapor also may exert an anti-staling effect.

In-package ethanol vapor generators are also used in Japan to extend the mold-free shelf life of bread, doughnuts, and sponge cake stored at ambient temperature. Mold-free shelf life of Maderia cake has been extended by Ethicap® sachet for up to three weeks at ambient temperature and without adverse effect on sensory characteristics. Ethicap® has also been used as a means for further extending the shelf life of modified-atmosphere packaged apple turnovers.

Freund's research indicates that some of the ethanol is lost by permeation through the food package material. Typical package films such as low-density polyethylene have an ethanol permeability of 20 to 30 g/m^2 day at 30°C (86°F). Polyvinyl chloride-polypropylene-polyethylene laminations (not used in the United States), according to Freund, have an ethanol permeation of about 1 to 2 g/m^2 day.

In a notable study in this area, objectives were to determine the effect of water activity (a_w) on vaporization of ethanol from Ethicap® into package headspace, to determine the effect of a_w and ethanol vapor on growth of *S. cerevisiae*, and to determine the suitability of ethanol vapor for shelf life extension of apple turnovers (Smith, 1988).

A preliminary study was done on a model system to determine the effect of water activity on vaporization of ethanol from different sizes of Ethicap® into the package headspace. The results demonstrated that water activity has a pronounced effect on both the vaporization of ethanol into, and level of ethanol remaining in, the package headspace.

A combination of water activity and ethanol vapor has also been shown to be effective in controlling growth and fermentation of *S. cerevisiae*, the major spoilage isolate of modified-atmosphere packaged apple turnovers. A reduction in water activity alone was demonstrated to be of limited practical use to control yeast growth and fermentation in apple turnovers.

Yeast growth and fermentation, however, could be controlled at higher a_w through a combination of a_w and ethanol vapor. Growth was completely inhibited at a_w of 0.9 and 1 52% headspace ethanol. Yeast growth and carbon dioxide production were also completely suppressed in all samples adjusted to a_w of 0.85 and packaged under an initial headspace ethanol of approximately 0.56% to 1.15%. These studies show that ethanol vapor exerts an antimicrobial effect on yeast growth and that the concentration of headspace ethanol required for inhibition depends on both the amount of Ethicap® used and the a_w of the system. Studies with apple turnovers with a_w of 0.93 and packaged under various atmospheres indicate that ethanol vapor can be used effectively to extend shelf life and to retain quality of these products. Yeast counts in air- and gas-packaged apple turnovers increased from zero after day one to approximately 106 after 21 days storage at 21°C. However, for product packaged under ethanol vapor, either alone or in conjunction with modified atmosphere packaging, yeast growth was completely inhibited. For the product packaged with Ethicap®, carbon dioxide production was completely inhibited and all packages appeared normal at the end of the 21-day storage period.

The research demonstrated that a longer shelf life of apple turnovers is possible by packaging the product with Ethicap® either alone or in conjunction with modified atmosphere packaging The advantages of using Ethicap® as a preservative were:

- Ethanol vapor can be generated without spraying ethanol solutions directly onto products prior to packaging.
- Sachets can be conveniently removed from packages and discarded at the end of the storage period.
- Encapsulated ethanol eliminates the need to use other preservatives such as benzoic acid or sorbic acid to control yeast growth.
- It does not impart any odor to the product.
- It is relatively inexpensive.

A disadvantage of using ethanol vapor for shelf-life extension of apple turnovers or other products is the absorption from package headspace by the product. The concentration of ethanol found in apple turnovers (1.45% to 1.52%), however, is within the maximum level of 2% by product weight permitted by regulatory authorities in the United States when ethanol is sprayed onto pizza crusts prior to final baking. Preliminary studies have shown that the ethanol content of apple turnovers can be reduced to 0.1% by heating the product at 190°C (374°F) for 10 minutes prior to consumption, a not-unusual food-preparation procedure. Therefore, while a longer shelf life may be possible by packaging products with Ethicap®, its use as a preservative may be limited at present to "brown and serve"-type baked products.

Japan's Morinanga Company patented an adhesive-backed, ethanol-containing film that can be taped on the inside of a package. The film contains ethanol encapsulated on the surface, which then evaporates into the package airspace to impart a preservative effect.

A Dutch patent involves adding an ethanol-containing gel into the package headspace.

Mitsubishi has a patent on a sachet that contains encapsulated ethanol as well as glucose, ascorbic acid, a phenolic compound, and an iron salt. Thus the sachet contents scavenge oxygen as well as emitting ethanol.

Asahi Denka Kogyo of Japan also patented a sachet that emits ethanol from a mixture encapsulated with cyclodextrins.

Encapsulated ethanol appears to have sufficient promise as an in-package preservative to warrant further investigation for soft bakery goods.

WASABI DERIVATIVE

Japan's Sekisui Jushi has developed an antibacterial food-packaging material based on a compound extracted from Japanese horseradish, wasabi. Polyethylene film containing horseradish-derived allylisothiocyanate (AIT) is enclosed in a cyclic oligo saccharide. The material is claimed to suppress the proliferation of bacteria and fungi on food surfaces, thereby prolonging shelf life.

Antibacterial sheets intended to be placed between layers of food, such as in lunch boxes or ready-to-eat foods, were originally designed for industrial use but are now available for retail consumers in Japan. Sekisui Kaseihin Kogyo is offering an antibiotic "freshness-retaining" sheet containing horse-radish extract for use in lunch boxes. This product functions to retain product freshness because of the antimicrobial properties of horseradish, moisture absorption, and the extra-moisture-holding properties of paper. After the sheet is placed on a lunch box with the printed part up, the box is closed. The antibiotic agent then spreads by means of the moisture present. The sheet is made of ink/antibiotic layer/water-absorbing paper/polyethylene with perforations. Extra water is absorbed by means of the perforations in the polyethylene. Another type of sheet developed by Japan's Lintec is made of wasabi, a derivative of Japanese horseradish, and mustard. The joint venture of Lintec and wrapping-paper maker Fuji Techno reports that both ingredients are effective in preventing bacterial and mold growth.

The antimicrobial properties of allylisothiocyanate (AIT) have also been applied by Japan's Green Cross, which has developed a mustard extract called WasaOuro®. WasaOuro® is the collective name of preparations derived from mustard, radish, or horseradish extract that contain the same type of oil as wasabi, a flavoring agent. The company claims to have completed the filing to the regulatory authorities for "mustard extract" for use as "nonsynthetic food additives."

The main ingredient of this mustard oil is AIT. This substance has already been approved for use (in Japan) as a food additive to deliver wasabi flavoring in foods. It has also been known to have strong antibacterial and antifungal effects, but has an irritating odor and acrid or pungent taste. Thus, when used as an antimicrobial, the compound is generally used indirectly in the package material. The antimicrobial effect is significantly increased when the AIT is in gaseous form rather than in direct contact with the food. WasaOuro® was engineered to control the emission of the volatile ingredient of mustard oil. In this way, the maximum antimicrobial effect is achieved with minimum absorption of AIT and with minimal effect on the flavor of the food. Unpublished research indicates that the odor of wasabi permeates food products at concentrations at which its antimicrobial effect is demonstrable.

ADDITIONAL ANTIMICROBIALS

Hinokitiol, or beta thujapricin, derived from cypress bark is frequently cited in trade literature as an antimicrobial agent incorporated into package structures in Japan.

Japan's Takex has introduced a product line of antibacterial items that use bamboo extract. Years ago, parts of bamboo trees were used for wrapping.

Takex has become one of the first to expand on this practice to use the natural bacteria-killing characteristics of bamboo in their antibacterial products. The main ingredient of TakeGuard products is taka kinon extract. According to the maker, the TakeGuard products can be used safely with food products.

U.S. Department of Agriculture (USDA) scientists have identified five natural berry aroma compounds that block the growth of molds. One compound in particular, 2-nonanone, exists in most fruits and is especially promising because of its fruity floral aroma, chemical stability, and low cost (less than U.S.$0.01 for enough to treat a quart of berries, excluding application costs). USDA scientists screened several hundred compounds that partially make up berry aroma and examined 15 chemicals in detail. They sealed small amounts of the compounds in jars containing berries inoculated with *Botrytis cinerea*. Five of the fifteen compounds prevented mold growth for at least a week at 10 °C. Researchers are testing to incorporate the 2-nonanone into a packaging material for berry application. Although they have worked only with berries to date, the compounds may exist in other fruits and could be packaged with susceptible fruits to keep them fresher longer.

CHITIN AND CHITOSAN

Chitin is a naturally occurring acetylated aminopolysaccharide (Porter et al., 1995). It forms the outer protective coating for crustaceans and insects in a covalently bound network with proteins and dihydroxy phenylalanine, and with some metal and carotenoid contaminations. Its deacetylated derivative, chitosan, is a polycationic long-chain biopolymer with a natural affinity for the normally negatively charged biological membranes.

Chitosan films have low oxygen permeability (i.e., are good gas barriers) (Mayer et al., 1989), and chitosan laurate films have low water permeability (i.e., are good moisture barriers) (Pennisi, 1992). Chitosan is a strong film former; in addition, chitosan can form polymer blends with other candidate polysaccharides.

Although now sold to consumers for fat reduction, neither chitin or chitosan has yet been cleared by the FDA for human consumption, but many foods contain much of it in the form of natural residues. Some possible applications of chitin and chitosan that have been suggested in the literature or are in current use are briefly discussed below.

Many workers have noted that chitosan is inhibitory to certain strains of bacteria and fungi (Knorr, 1991; Popper and Knorr, 1990). Chelation of essential metals such as zinc, necessary for growth, is a possible cause (Cuera et al., 1991), but the agglomeration capacity of the polycation for the anionic microbes is also possible. Ralston et al. (1964) reported that chitosan inhibited baker's yeast fermentation by preventing glucose uptake.

N-Carboxymethyl chitosan gave 90% inhibition of iron-activated autoxidation. This is regarded as a chelation action, and *N*-carboxymethyl chitosan is a potent metal chelator.

The U.S. Army Natick RD&E Center (now the Natick Soldier Center) studied three food applications of the antioxidant action of chitosan:

- in freeze-dried, unprocessed whole milk
- in freeze-dried, buffer-diluted heavy cream
- in washed surimi

In each case, the drying procedure tends to break up fat globule membranes and expose autoxidizable milk fat to catalysts like metal and enzymes. In all cases, the chelating action of chitosan exerted an antioxidant effect.

Chitosan was shown to be comparable to propyl gallate in dried milk in antioxidant effectiveness.

The antimicrobial spectrum of chitosan laurate was evaluated using a "screening" assay technique. The matrix of test organisms selected for inclusion were 32 strains of known or potential foodborne pathogens and spoilage organisms. A sample of powdered chitosan was dissolved in 0.1% acetic acid under conditions of mixing and mild heat. With the exception of the psychrotrophic organisms, which were incubated at 30°C, all the strains were incubated at 35°C. Three levels of inhibition were observed among the test organisms (Table 26).

To assess the antimicrobial efficacy of chitosan in foods, beef frankfurters coated with calcium alginate gel containing 0 to 2% chitosan solution were

TABLE 26. **Relative Sensitivity of Selected Bacteria to Chitosan Lactate.**

Complete Clearing/ Inhibition	Strong Inhibition	Partial Inhibition
Bacillus cereus NRRL 569	*Escherichia coli* O157: H7 HC760	*Salmonella typhimurium* 83
Shigella sonnei ATCC 25932	*Shigella flaxneri* ATCC 12022	*S typhimurium* ATCC 14028
Listeria monocytogenes Scott A	*Pseudomonas lundensis* 385	*S enteritidis* 56
Staphylococcus aureus	*Citrobacter freundii* ATCC8090	*S heidelberg* 3432-2
S epidermidis ATCC 12228	*Enterococcus faecalis* ATCC 19433	
Proteus vulgaris ATCC 13315		
Yersinia enterocolitica P106Y		

Source: Porter et al , 1995

inoculated with *Listeria monocytogenes* and incubated at 25°C. No inhibition of *Listeria* was observed (Porter et al., 1995).

ANTIMICROBIAL PADS

A Kimberly Clark patent (Hansen et al., 1989) describes a pad that absorbs exudate or purge from packaged food products, particularly fresh meat and poultry, and inhibits propagation of foodborne pathogens. The pad contains an effective amount of an antimicrobial composition of an acid having the structure $R-COOH$, where R is one of a group of various lower alkyls, lower alkenyls, or phenyl and substituted phenyls. A mixture of citric acid, malic acid, and sodium lauryl sulfate in a ratio of 10:5:2 by weight is dispersed within the absorbent medium of the pad at a concentration of 0.4% to 0.5% of the exudate weight.

ANTIMICROBIAL PACKAGES

Japan's Kachi Seibukuro offers a range of antibacterial film bag styles. The film exhibits germ-killing action by inhibiting the proliferation of such bacteria as *Salmonella*. The film will not dissolve in water or solutions, and the compounds responsible for the microbial-killing properties will not evaporate. The film is claimed to meet Japanese government safety guidelines.

Japan's Daiya Foods has developed an antibacterial egg pack and food container that exhibits microbicidal properties through the use of fine ceramics and thus does not actually touch the foods inside the packages. Freshly laid eggs are usually given a microbicidal wash and then distributed. Bacteria may remain and proliferate during distribution. The egg pack from Daiya Foods curtails the spread of microorganisms. The bacteriocidal properties are semi-permanent. The cost of an egg pack is about U.S.$0.03 higher than regular egg packs and about U.S.$0.04 higher than specialty egg packs.

NON-FOOD ANTIMICROBIAL DELIVERY SYSTEMS

An antibacterial wound dressing that automatically releases antibiotic when the wound has become infected has been developed by a group from Japan's Nara Advanced Institute of Science and Technology, Kyoto University, and Kuraray Company. Wound dressings are used on burn victims and in other cases of severe injury to cover and protect the wound and to promote healing. Bacterial infections can become life-threatening to patients weakened from a serious injury, and so these conventional wound dressings contain antibiotic medicines. However, conventional dressings continually release antibiotic into the wound, and over long periods of time there is a risk that some bacteria will develop resistance to the drug.

The new antibacterial dressing is a type of so-called intelligent material that only releases its antibiotic when the wound becomes infected with bacteria. The antibiotic, in this case gentamicin, is fixed to a thin film of a biocompatible polymer material via a peptide molecule. When the wound becomes infected by bacteria such as *Staphylococcus* or *Pseudomonas,* peptides are severed to release the drug. The researchers claim they have already confirmed with tests using mice that the antibiotic is only released when the wound becomes infected.

In India, studies were conducted on the effectiveness of antimicrobial agents hisaplin and potassium sorbate and package materials aluminum foil, polyethylene, and parchment paper on microbial quality of a product called khoa during storage at 37°C (98.5°F) and 5°C (41°F). Results showed that counts of mesophilic aerobes and fungi in khoa were reduced by the incorporation of 0.30% potassium sorbate and by packaging in aluminum foil (Ghosh et al., 1973, 1977).

ANTIMICROBIALS IN PAPER

A Russian study of fungistatic agents suitable for use in fruit and vegetable packaging papers tested sodium *o*-phenylphenate (Preventol ON) and *o*-phenylphenol propionic acid. The test microorganisms were *Rhizopus nigricans, Botrytis cinerea, Penicillium expansum, Monilia fructigena,* and *Trichothecium roseum.* The agents had about the same activity, inhibitory to all fungi except *Rhizopus nigricans* in amounts of 0.5%, and effective against *Rhizopus* in an amount of 1%. Preventol ON was recommended for further study because of its higher solubility and maintenance of the fungicidal action on storage of treated paper.

One patent for a bacteriocide for paperboard or corrugated fiberboard describes a solution containing at least two quaternary ammonium salts of which at least one possesses a higher alkyl group with 8 to 12 carbon atoms. The antimicrobial agent can be applied to the material by spraying.

A Japanese Dai Nippon patent describes fungus-resistant paper for packaging fruit prepared by using 0.08% to 10.0% of an acid adjunct of 1,17-diaguanidino-9-azaheptadecane as a fungicide. The paper contains 25 ppm of the compound. An antiseptic paper for packaging was prepared from rayon fiber webs treated with copper sulfate A viscose solution was spun to form fibers containing 5% carbon disulfide, the fibers were formed into a web, and the web was treated with 2% aqueous copper sulfate to form antiseptic paper.

Japan's Matsushita (Yoshida et al., 1993) developed paper pulp slurried in water, to which a paper-strengthening agent such as polyamidepolyamine-epichlorohydrin condensate as well as other compounds are added. Calcium carbonate or calcium phosphate treated with a silver, copper, or zinc salt, e.g., calcium carbonate treated with copper sulfate or zinc oxide, are added as an antimicrobial agent. The mixture is incorporated into a sheet of paper that is antimicrobial and used as the inner liner for corrugated fiberboard in cases.

In Italy, bactericidal and fungicidal properties of trialkyltin compounds were investigated relative to the protection of paperboard containers against microorganisms. The inhibition of bacterial growth resulting from organotin compounds was tested on *Escherichia coli* and *Staphylococcus pyogenes* var. *aureus*. Tripropyltin compounds showed better antibacterial action than did tributyltin compounds. The acidic group of the tested organotin compounds (benzoate, salicylate, etc.) did not show any significant influence on the bactericidal action. The inhibition of growth of *Aspergillus niger* by Lastanox P (20% bis-tributyltin oxide + chlorinated PhOH + surfactants), Lastanox T (20% bis-tributyltin oxide + surfactants), tributyltin oxide (TBTO), Estabex L (a dialkyltin preparation) as compared to that of sodium pentachlorophenate was also investigated. The best fungicidal effect was observed with TBTO. A high affinity of organotin compounds toward cellulose was observed (Bomar, 1968).

Japan's Mizasawa has a patent in which amorphous tectosilicate (pore diameter, 20 to 300 Å; cation exchange volume 2 to 100 meq/100 g), which is composed of silicon dioxide:aluminum oxide:sodium oxide in a ratio of 1:0.04 to 0 4:0.01 to 0.1, is slurried in water. An aqueous antimicrobial metal salt (0.05% to 10%) such as silver nitrate is added to the mixture. An organic antimicrobial compound (0.5 to 15 wt.%) such as cetylpyridinium chloride is added to the dry precipitate. The antimicrobial composition is added to paper for use in wrapping fresh food (Suzuki et al., 1991).

In a Japanese patent, paper is coated with a dispersed mixture of a silver, copper, lead, or tin salt of an *N*-(12 to 18 carbon) acyloylamino acid (0.1 to 1 wt.%) such as silver *N*-stearoyl-L-glutamate and a binder such as polyvinyl alcohol in water. The amount of the amino acid salt applied is 0.01 to 1 g/m^2. The antimicrobial paper is used to wrap food.

Japan's Dai Nippon has a patent on paper bags for protection of fruits from pathogenic microorganisms. The paper is impregnated with salts of microbicidal (9 or more carbon) guanidine compounds containing lipophilic groups. One part by weight guazatine tris(dodecylbenzene sulfonate) and 1 part by weight carbon are incorporated at levels of 0.38 part of this powder into 100 parts paraffin at 100°C. A paper is immersed in this melt. A fruit-protecting bag prepared from this sheet is not permeable to the fungus *Alternaria kikuchiana*.

Japan's Yokata developed paper pulp slurried in water to which radioactive mineral powder containing 0.5 to 2 wt.% thorium oxide, silver-containing mineral powder (silver content, higher than 0.005 wt.%), and/or zinc oxide powder singly or in combination are added. The mixture is incorporated into paper sheets used to wrap fresh food.

Japan's Oji Paper has a patent on a paper pulp slurry to which is added a metal salt of an oxyquinoline derivative (0.05 to 5 wt.% of pulp) such as a copper salt of 8-oxyquinoline, a humic acid salt (0.01 to 1 wt.% of pulp) such as a sodium or ammonium salt, and a water-soluble multivalent metal salt (1 to 10 wt.% of pulp) such as aluminum sulfate. The mixture is converted to a sheet in which the antimicrobial agent is fixed in high yield. The sheet obtained is used for wrapping of fruits.

In India, an effective fungistatic wrapper can reportedly be made economically by coating grease-proof paper with an aqueous dispersion of sorbic acid and an antioxidant (viz., Embanox-6) in 2% carboxymethyl cellulose solution. Foods can be preserved for a minimum of 10 days by wrapping them with the sorbic acid–treated paper (while they are still hot, i.e , four minutes or less after their preparation) and then enclosing them in a polyethylene bag

INDIVIDUAL PRODUCT SEALING

In Israel, shrink-film seal packaging developed by the Volcani Institute offers a means of providing fungicides to individual fruits by incorporating the fungicide in the packaging film. Film containing the fungicide imazalil, (±)-1-[β-(allyloxy)-2,4-dichlorophenethyl] imidazole, markedly reduced the decay of Shamouti oranges inoculated with *Penicillium digitatum* if sealing was performed immediately following the inoculation and only slightly if sealing was delayed 24 hours. The film could serve as a reservoir releasing the fungicide gradually, so that its antifungal activity may replace that previously provided by waxing without the risk of excessive toxic residues (Ben-Yehoshua et al., 1987). Sulfur dioxide release into packages of table grapes can be achieved by use of microporous pads containing sodium metabisulfite. This chemical is hydrolyzed by the action of water vapor in the headspace of the carton liner.

PHOTOCATALYSTS

First developed in Japan as antimicrobials, photocatalysts, in the presence of light, remove "dirt," odor, or organic molecules that come in contact with the surface on which the photocatalysts reside. These photocatalysts also have antibacterial effects. A photocatalyst derives catalytic activity by absorbing energy from a light source. When the photocatalyst is irradiated with ultraviolet radiation, highly reactive oxygen free radicals are generated in the surrounding area to bring about the decomposition of organic molecules.

Zinc oxide, lead oxide, and silver salts have photocatalytic properties, but the most promising material is titanium dioxide, already widely used in paints, foods, and cosmetics. Although the compound is normally white, it becomes transparent when ground to a fine powder. Results to date suggest that titanium dioxide can be mixed in with just about any type of material to help destroy air pollutants and prevent items from getting dirty.

A partial drawback to photocatalysts is that since plastics are made from organic molecules, they themselves would degrade in the presence of a photocatalyst. However, Japan's Yamamura Glass developed a means to plate titanium dioxide inside silicon-oxide microcapsules. Encapsulated in this manner, the titanium dioxide cannot come into contact and degrade the plastic; meanwhile, odor and other molecules small enough to slip through the numerous nano-scale pores of the microcapsule encounter the photocatalyst and are destroyed. Unfortunately, however, existing photocatalysts can decompose any organic molecule that comes near. It would be far more efficient if photocatalysts would have the ability to selectively degrade certain molecules, such as those of odors or pollutants.

A new photocatalyst that can capture bacteria and break them apart has been developed by a group from the National Industrial Research Institute of Nagoya, Japan. Combining the microbial-breaking properties of titanium dioxide with the adsorbency of apatite ($Ca_5F(PO_4)_3$), the new composite functions continuously to sterilize any surface to which it is applied. The blend may be mixed into plastics. Titanium dioxide is a photocatalyst that when exposed to light has the ability to decompose organic molecules, i.e., it can kill bacteria and other microorganisms. It cannot, however, attract bacteria, and so it only works on organic molecules with which it makes direct contact. Apatite is a ceramic that absorbs proteins and so can capture bacteria. The new composite combines the benefits of both materials by coating powders or thin films of titanium dioxide with a micron-thick layer of apatite. The apatite adsorbs proteins that project from the surface of the bacteria, and the titanium dioxide then proceeds to tear their organic molecules apart, killing the bacteria in the process In experiments using *E. coli*, the composite material provided 100% sterilization.

Development is proceeding on other photocatalysts like silver and lead oxide to boost the efficiency of the decomposition reactions.

MATHEMATICAL MODELING OF ANTIMICROBIAL FOOD PACKAGING[1]

Foods, unlike durable goods such as electronics and furniture, are usually perishable products and heterogeneous mixtures. It is not easy to create mathematical models for the physical and chemical properties of foods to predict their behavior during processing. In addition, foods have safety aspects and a relatively very short shelf life. Therefore, food packaging is quite different from the packaging of durable products.

[1]Reproduced with permission of Institute of Food Technologists This section appeared as a paper by Jung H Han, entitled "Antimicrobial Food Packaging," in *Food Technology*, March 2000, 54(3):56–65

Food-packaging materials used to provide only barrier and protective functions. However, various kinds of active substances can now be incorporated into the packaging material to improve its functionality and give it new or extra functions. Such active packaging technologies are designed to extend the shelf life of foods while maintaining their nutritional quality and safety.

Active packaging technologies involve interactions among the food, the packaging material, and the internal gaseous atmosphere (Labuza and Breene, 1988). The extra functions they provide include oxygen scavenging, antimicrobial activity, moisture scavenging, ethylene scavenging, ethanol emitting, and so on.

Floros et al. (1997) reviewed existing active packaging products and patents. The most promising active packaging systems are oxygen-scavenging systems—which absorb oxygen gas in the package and prevent rancidity of foods (Rooney, 1981) and are being developed as forms of sachets or polymer additives (Rooney, 1996)—and antimicrobial systems. This section focuses on antimicrobial systems.

ANTIMICROBIAL FOOD PACKAGING

When antimicrobial agents are incorporated into a polymer, the material limits or prevents microbial growth. This application could be used for foods effectively not only in the form of films but also as containers and utensils. Food package materials may obtain antimicrobial activity by common antimicrobial substances, radiation, or gas emission/flushing. Radiation methods may include using radioactive materials, laser-excited materials, ultraviolet-exposed films, or far-infrared-emitting ceramic powders. However, universal irradiation sterilization of food package materials is not yet permitted by the FDA.

Gas emission/flushing controls mold growth. For examples, berries and grapes are stored in produce boxes, palletized, and stretch wrapped, then flushed with sulfite to prevent fungal spoilage. It is easy to use these types of bulk gas flushing and controlled-/modified-atmosphere technologies. However, there is no commercial material that contains or releases sterilizing gases such as sulfite. Sachet systems have been used to control the gas composition inside a package; e.g., an ethanol-vapor-generating sachet can inhibit mold growth on bakery products (Smith et al., 1987).

An oxygen-scavenging system absorbs oxygen gas in the package and prevents growth of aerobic microorganisms, especially mold, as well as oxidation of food components.

Table 27 reviews some typical applications of antimicrobial packaging systems.

Antimicrobial packaging materials have to extend the lag period and reduce the growth rate of microorganisms to prolong shelf life and maintain food safety. They have to reduce microbial growth of nonsterile foods or maintain the

TABLE 27. Applications of Antimicrobial Food Packaging.

Antimicrobial Agent	Packaging Material[1]	Food	Reference
Organic Acid			
Potassium sorbate	LDPE	Cheese	Han (1996)
	LDPE	Culture media	Han and Floros (1997)
	MC/palmitic acid	Culture media	Rico-Pena and Torres (1991)
	MC/HPMC/fatty acid	Culture media	Vojdani and Torres (1990)
	MC/chitosan	Culture media	Chen et al. (1996)
	Starch/glycerol	Chicken breast	Baron and Sumner (1993)
Calcium sorbate	CMC/paper	Bread	Ghosh et al. (1973, 1977)
Propionic acid	Chitosan	Water	Ouattara et al. (1999)
Acetic acid	Chitosan	Water	Ouattara et al. (1999)
Benzoic acid	PE-co-MA	Culture media	Weng et al. (1997)
Sodium benzoate	MC/chitosan	Culture media	Chen et al. (1996)
Sorbic acid anhydride	PE	Culture media	Weng and Chen (1997), Weng and Hotchkiss (1993)
Benzoic acid anhydride	PE	Fish fillet	Huang et al. (1997)
Fungicide/Bacteriocin			
Benomyl	Ionomer	Culture media	Halek and Garg (1989)
Imazalil	LDPE	Bell pepper	Miller et al. (1984)
	LDPE	Cheese	Weng and Hotchkiss (1992)
Nisin (peptide)	Silicon coating	Culture media	Daeschel et al. (1992)
	SPI, corn zein films	Culture media	Padgett et al. (1998)

Peptide/Protein/Enzyme			
Lysozyme	PVOH, nylon, cellulose acetate SPI film, corn zein films	Culture media	Appendini and Hotchkiss (1996)
		Culture media	Padgett et al. (1998)
Glucose oxidase	Alginate	Fish	Field et al. (1986)
Alcohol oxidase	—	—	Brody and Budny (1995)[2]
Alcohol/Thiol			
Ethanol	Silica gel sachet	Culture media	Shapero et al. (1978)
	Silicon oxide sachet (Ethicap®)	Bakery	Smith et al. (1987)
Hinokitiol	Cyclodextrin/plastic (Seiwa™)	—	Gontard (1997)[2]
Gas			
CO_2	Calcium hydroxide sachet	Coffee	Labuza (1990)[2]
SO_2	Sodium metabisulfite	Grape	Gontard (1997)[2]
Other			
UV irradiation	Nylons	Culture media	Paik and Kelley (1995), Hagelstein et al. (1995)
Silver zeolite	LDPE	Culture media	Ishitani (1995)
Grapefruit seed extract	LDPE	Lettuce, soybean sprouts	Lee et al. (1998)

[1]LDPE = low-density polyethylene; MC = methyl cellulose; HPMC = hydroxypropyl MC; CMC = carboxyl MC; PE = polyethylene; MA = methacrylic acid; SPI = soy protein isolate; PVOH = polyvinyl alcohol; BHT = butylated hydroxy toluene; HDPE = high-density PE.
[2]From review articles without experimental data.

stability of pasteurized foods without post-contamination. If the packaging materials have self-sterilizing ability because of their own antimicrobial activity, they may eliminate chemical sterilization of packages using peroxide and simplify the aseptic packaging process (Hotchkiss, 1977). The self-sterilizing materials could be widely applied for clinical uses in hospitals, biological labware, biotechnology equipment, and biomedical devices, as well as food packaging.

ANTIMICROBIAL SUBSTANCES

A chemical preservative can be incorporated into a packaging material to add antimicrobial activity to it. For example, preservative-releasing films provide antimicrobial activity by releasing the preservative at a controlled rate. Also, oxygen absorbents reduce headspace oxygen and thus partially protect food against aerobic spoilage such as mold growth (Smith et al., 1990).

Common antimicrobial chemicals for food products are such preservatives as organic acids and their salts, sulfites, nitrites, antibiotics, and alcohols (Table 27). For example, sorbic acid and its potassium salts have been studied as preservatives for the packaging of cheese products. They were mixed into a wax layer for natural cheese (Melnick and Luckmann, 1954a, 1954b, Melnick et al., 1954; Smith and Rollin, 1954a, 1954b); wet wax coating on packaging paper (Ghosh et al., 1973, 1977), and edible protein coating on intermediate-moisture foods (Torres et al., 1985). However, the release rate and migration profile of antimicrobial agents in these applications were not specifically controlled. The antimicrobial mechanism/kinetics and the controlled-release profile of potassium sorbate from low density polyethylene (LDPE) film into cheeses were examined and mathematically simulated by Han (1996).

Other attempts at incorporating chemicals into plastics for use as antimicrobial packaging films involved antimycotics (fungicides) and antibiotics. Imazalil was used as the active substance and was chemically coupled to plastic films to delay the growth of molds. A shrink-wrapping film of imazalil in LDPE for use on peppers (Miller et al., 1984) and cheddar cheese (Weng and Hotchkiss, 1992) and an imazalil-bound ionomer film (Halek and Garg, 1989) showed antifungal properties and controlled the contamination of cheese and peppers.

Other chemicals, gases, enzymes, and natural components have been tested as preservatives or sterilizing agents. The agents tested include propionic acid, peroxide, ozone, chlorine oxide, eugenol, cinnamaldehyde, allyl isothiocyanate, lysozyme, nisin, and EDTA. These antimicrobial agents may be incorporated into packaging materials. Compared to hydrogen peroxide, ozone, and chlorine oxide, other natural agents may have more advantages for the antimicrobial packaging system, as they are considered edible components.

Biodegradable polymers are currently being studied as edible coatings or film materials (Krochta and De Mulder-Johnston, 1997). Padgett et al. (1998)

demonstrated the antimicrobial activity of lysozyme and nisin in soy protein isolate films and corn zein films. Use of edible films or coatings incorporating food preservatives and natural antimicrobial agents may become popular areas of packaging research.

ANTIMICROBIAL PACKAGING SYSTEMS

Most food packaging systems consist of the packaging material, the food, and the headspace in the package. If the void volume of solid food products is assumed as a kind of headspace, most food packaging systems represent either a package/food system or a package/headspace/food system (Figure 3).

A package/food system has a package-contacted food product or a low-viscosity or liquid food without headspace. Individually wrapped cheese, deli products, and aseptic brick packages are good examples. Diffusion between the packaging material and the food and partitioning at the interface are the main migration phenomena involved in this system. Antimicrobial agents may be incorporated into the packaging materials initially and migrate into the food through diffusion and partitioning.

Examples of a package/headspace/food system are flexible packages, bottles, cans, cups, and cartons. Evaporation or equilibrated distribution of a substance among the headspace, packaging material, and/or food has to be con-

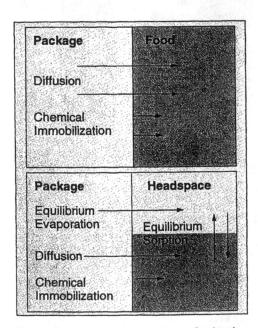

Figure 3 Food packaging systems and migration phenomena

sidered as a part of main migration mechanisms to estimate the interfacial distribution of the substance Compared to a nonvolatile substance, which can only migrate through the contact area between the package and the food, a volatile substance can migrate through the headspace and air gaps between the package and the food.

Besides diffusion and equilibrated sorption, some antimicrobial packaging uses covalently immobilized antibiotics or fungicides. This case utilizes surface suppression of microbial growth by immobilization of the non-food-grade antimicrobial substance without diffusional mass transfer. Figure 4 shows the mass transfer phenomena of an active substance with different applications.

Control of the release rates and migration amounts of antimicrobial substances from food packaging is very important. Biochemical factors affecting the mass transfer characteristics of antimicrobial substances include antimicrobial activity and the mechanism/kinetics of selected substances to target microorganisms. Studies on these problems provide information on the exact

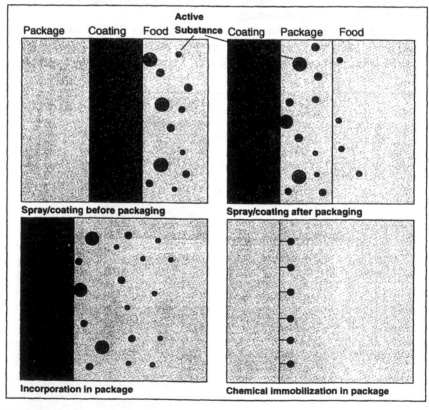

Figure 4 Migration of active substance

amount and release rate of the antimicrobial substances required to achieve a given effect. The release kinetics have to be designed to control the growth kinetics and maintain the antimicrobial concentration above the critical inhibitory concentration.

DESIGNING THE PACKAGING SYSTEM

A number of factors affect the design or modeling of an antimicrobial film or package.

Casting Process Conditions and Residual Antimicrobial Activity

Antimicrobial activity of an incorporated active substance may deteriorate during casting (film or container), converting, and/or storage and distribution of the packaging material. The residual antimicrobial activity is the effective activity of the antimicrobial agents utilized for the antimicrobial packaging of final films after the casting and converting processes.

Most activity deterioration may occur during the extrusion process using high temperature, shearing force, and pressure. During extrusion of plastic resins, the high pressure and temperature conditions in the extruder affect the chemical stability of incorporated antimicrobial substances and reduce their residual antimicrobial activity (Han and Floros, 1999). The temperature, pressure, and residence time in the extruder have to be quantified mathematically to predict the residual antimicrobial activity. The effects of converting operations such as lamination, printing, drying, and storage processes and the effects of adhesives and solvents should also be characterized quantitatively. Besides chemical degradation, loss of volatile antimicrobial compounds is also a reason for antimicrobial activity loss during casting (extrusion or coating) and storage of the packaging materials.

Characteristics of Antimicrobial Substances and Foods

The growth-inhibition mechanism and kinetics are the first factors to be considered in designing the antimicrobial packaging system. The mathematical model of microbial growth can be built up from the inhibition kinetics and mechanism. The release kinetics of antimicrobial agents have to be designed to maintain the concentration above the critical inhibitory concentration with respect to the growth kinetic studies.

Because foods have different chemical and biological characteristics such as pH, water activity, carbon source, nitrogen source, partial pressure of oxygen, and temperature, they provide different environmental conditions to microorganisms and included antimicrobial agents. For example, the pH of food affects the microflora and growth rate of target microorganisms and alters the ionization (dissociation/association) of most active chemicals, which could change the antimicrobial activity. Water activity also alters the antimicrobial activity and chemical stability of incorporated active substances as well as microflora. Oxygen in the package headspace can be utilized by aerobic microorganisms, and the oxygen permeability of the packaging materials can alter the headspace oxygen concentration. Therefore, it is essential to study the pH and water activity of the food, oxygen permeability of the packaging material, and microbial profile to design antimicrobial packaging systems (Smith et al, 1989).

Storage Temperature

Storage temperature can change the antimicrobial activity of the chemical preservatives. The temperature conditions during production and distribution have to be recorded to determine the heat effect on the residual antimicrobial activity. The combination of heat treatment, storage temperature, and preservative may have a synergistic effect because of the increased heat and chemical sensitivity of microorganisms. Volatile chemicals may be affected by the partial pressure in the headspace and the characteristic equilibrium constant among the headspace, packaging materials, and foods with respect to the storage temperature.

Mass Transfer Coefficients

The simplest system is diffusional release of active substances from the package into the food. An antimicrobial packaging system has at least a two-layer structure: the packaging material and the food. If laminated multilayer materials are used, the system has a structure with more than three layers: outer layer, inner layer, and the food. In laminated multilayer materials, the diffusivity of each layer affects the overall diffusion profile of an active substance. The diffusivity in food also affects the overall diffusion profile of the active substance, as well as the diffusivity in the packaging material.

Han (1996) suggested a multilayer structure for antimicrobial-release packaging systems: outer antimicrobial barrier layer (optional); antimicrobial-containing matrix layer; release-control layer; and food (Figure 5). The outer layer is a barrier layer to prevent loss of active substance to the environments and gives a unidirection of diffusion forward to the inner layer. The matrix layer contains active substances and has a very fast diffusion of the antimicrobials and a very large portion of amorphous structure to provide a space for the antimicrobials. The control layer is the key layer to control the initial lag period and the flux of penetration of the antimicrobials. It has a customized thickness and diffusivity with respect to the characteristics of microbial spoilage and food products. This multilayer design has advantages of easy

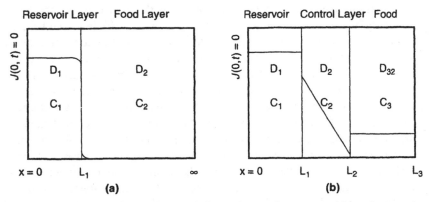

Figure 5 Multi-layer mass transfer models for food packaging systems: (a) two-layer semi-infinite model, and (b) three-layer finite model

customization by deciding on the thickness and material of the control layer to alter the diffusivity of antimicrobial agents. Selected control-layer film will be laminated to the premade antimicrobial base film consisting of an antimicrobial-matrix layer and an outer layer. Containers as well as films can be formed from the multilayer sheet of antimicrobial material by vacuum or press-molding processes. Papers as well as plastics can be used as antimicrobial package materials. Because paper has a porous structure, the antimicrobial agents plugged in the pores may improve the paper material's performance, such as water vapor and gas permeability, physical strength, optical properties, and surface properties, as well as the antimicrobial activity.

When a substance moves through a matrix, the driving force of the transfer is concentration difference. Therefore, diffusivity of the moving substrate is the characteristic constant in the specific matrix system. When the moving substance passes over the medium boundary to the other heterogeneous matrix, however, either the concentration profile of the moving substance shows a discontinuous gradient at the interface of two layers, such as a sudden jump or drop in concentration, or no transport occurs, with a zero value of flux. Therefore, distribution constants at the interface as well as the diffusivity should be considered as a significant mass transfer coefficient. The distribution constants may include the partition coefficient, adsorption/desorption constant, solubility, and/or evaporation rate constant. The concentration difference at the interface is caused by the chemical potential equilibrium between two layers. In terms of energy, the moving substance at the interface is equilibrated and has the same value of free energy despite nonidentical concentration. The substance can move to the higher-concentration area from the lower-concentration area to equilibrate the chemical potential.

The mass transfer Biot number (also known as the Sherwood number), the ratio of surface mass transfer coefficient to diffusion coefficient, can be used for modeling the overall mass transfer profile to predict the concentration of the moving substance at any position and time. Most mass transfer experiments intentionally use the turbulent circulation of outside fluids to discard the boundary film resistance. Because of this turbulence with a large mass transfer Biot number, the thin boundary film resistance may not work as a bottleneck of overall mass transfer in the experimental systems. This lumped-capacity (large-Biot-number) experimental system can be used very easily for mass transfer modeling and diffusivity determination. However, because of the artificial nature of the boundary conditions of the lumped-capacity model, such as forced convection, the model is not suitable for describing natural mass transfer phenomena, which usually have large boundary resistances with small mass transfer Biot numbers.

To calibrate and estimate the mass transfer Biot number, the surface mass transfer coefficients have to be included in the overall mass transfer model. For example, the partition coefficient simplifies mass transfer models because it represents the interfacial (boundary) concentration distribution. Addition of the partition coefficient to the mass transfer model makes a single general model disregarding Biot numbers of the system, compared to the various experimental models currently available with different levels of Biot numbers.

Physical Properties of Package Materials

When antimicrobial activity is added to package materials to reduce microbial spoilage, it may affect general physical properties and process/machinability of the package materials. General properties of package materials include mechanical properties such as tensile strength, elongation, burst strength, tearing resistance, stiffness, and physical properties such as oxygen (and other gas) permeability, water vapor permeability, wettability, water absorptiveness, grease resistance, brightness, haze, gloss, transparency, and others.

The performance of the package materials must be maintained with the addition of the active substances, even though the materials contain more heterogeneous formulations. In the case of plastics, the active agents are usually very-low-molecular-weight chemicals compared to the size of the polymeric structure and are added in small amounts. In a properly designed antimicrobial packaging system, the chemicals will position themselves in the amorphous structural regions of the polymeric structure and may not affect the mechanical strength of the polymeric packaging materials. Han and Floros (1997) reported no significant change in tensile properties before and after incorporating potassium sorbate into low density polyethylene film.

Considering the huge size of the amorphous area of the polymeric materials (or porous area of papers) to the relatively small size of the antimicrobial molecule, a large amount of the antimicrobial substance may be needed to show any effect on the tensile strength of the packaging materials. However, after the amorphous area of the polymers and the porous area of papers are saturated by a high concentration or large powder particles of antimicrobial chemicals, the tensile strength of the antimicrobial materials could be adversely affected.

Besides mechanical strength changes, incorporation of antimicrobial agents usually reduces optical properties of plastic films such as transparency. For example, the transparency of low density polyethylene films decreased with increasing potassium sorbate concentration (Han and Floros, 1997). This may result in serious disadvantage in using this antimicrobial plastic film for see-through packaging.

MASS TRANSFER MODELING OF ANTIMICROBIAL SUBSTANCES

The mathematical model for the diffusion has to be established to explain the release profile of an active substance from an antimicrobial packaging material into a food product. This will permit the estimation of accurate concentration patterns, provide the diffusion profile of real food packaging systems, and predict the period in which the antimicrobial concentration will be maintained above the critical inhibitory concentration in the packaged food.

Model Development

In laminated multilayer materials, the diffusivity of the antimicrobial agent in each layer simultaneously affects the overall diffusion profile in other layers. Therefore, diffusivities of each layer and the food have to be defined correctly to estimate the concentration distribution. Compared to the packaging material layer, the food layer has a higher diffusivity and extremely larger volume, so that a semi-infinite model in which the packaging layer has a thickness L and the food layer has infinite volume could be practical.

Figure 6 shows the boundary and initial conditions of mass transfer models. The two-layer model (package/food) is a good system for package-contacted solid foods and liquid or viscous foods. There are two approaches to explain mass transfer of the two-layer system. One approach is to use a migration model that explains total mass of diffusant from inside the packaging material into the food without a specific position variable x on the migration direction. The fraction migration M_t/M_∞ of the active substance is shown in Equation 1 with assumption of $D_1 << D_2$, where D_1 is the diffusivity of the packaging material and D_2 is the diffusivity the liquid food, which is very fast and can be ignored (Carslaw and Jaeger, 1947, 1959; Luicov, 1968; Crank, 1975; Cardarelli, 1976).

Equation 1

$$\frac{M_t}{M_\infty} = 1 - \sum_{n=1}^{\infty} \frac{8}{(2n-1)^2 \pi^2} \exp\left[-D(n-1/2)^2 \pi^2 v L_1^2\right]$$

at $0 \leq x \leq L_1$
at $L_1 \leq x$ during all t

$$C(x,0) = C_0$$
$$C(x,t) = 0$$

Equation 2

$$\frac{C_1}{C_0} = 1 - \frac{1}{1+m} \sum_{n=1}^{\infty}(-h)^{n-1}\left[\text{erfc}\frac{(2n-1)L_1 - x}{2(D_1 t)^{1/2}} + \text{erfc}\frac{(2n-1)L_1 + x}{2(D_1 t)^{1/2}}\right]$$

$$\frac{C_2}{C_0} = \frac{m}{1+m}\text{erfc}\frac{x - L_1}{2(D_2 t)^{1/2}} - \frac{m(1+h)}{1+m}\sum_{n=1}^{\infty}(-h)^{n-1}\text{erfc}\frac{x - L_1 + 2nL_1(D_2/D_1)^{1/2}}{2(D_2 t)^{1/2}}$$

at $0 \leq x \leq L_1$
at $L_1 \leq x \leq L_2$

$$C(x,0) = C_0$$
$$C(x,0) = 0$$

all $t \geq 0$

$$\frac{\partial C_1}{\partial x}\bigg|_{x=0} = 0$$

$$D_1\frac{\partial C_1}{\partial x} = D_2\frac{\partial C_2}{\partial x}\bigg|_{x=L_1} \quad \text{and} \quad C_2(L_1,t) = C_1(L_1,t) \quad \text{at } t > 0$$

Equation 3

$$\frac{C_1}{C_0} = 1 - \frac{K_{12}}{m + K_{12}} \sum_{n=1}^{\infty}(-h)^{n-1}\left[\text{erfc}\frac{(2n-1)L_1 - x}{2(D_1 t)^{1/2}} + \text{erfc}\frac{(2n-1)L_1 + x}{2(D_1 t)^{1/2}}\right]$$

$$\frac{C_2}{C_0} = \frac{mK_{12}}{m+K_{12}}\text{erfc}\frac{x-L_1}{2(D_2 t)^{1/2}} - \frac{mK_{12}(1+h)}{m+K_{12}}\sum_{n=1}^{\infty}(-h)^{n-1}\text{erfc}\frac{x - L_1 + 2nL_1(D_2/D_1)^{1/2}}{2(D_2 t)^{1/2}}$$

$$D_1\frac{\partial C_1}{\partial x} = D_2\frac{\partial C_2}{\partial x}\bigg|_{x=L_1} \quad \text{and} \quad C_2(L_1,t) = K_{12}C_1(L_1,t) \quad \text{at } t > 0$$

Equation 4

4a. Mass Transfer Mode

$$J = \frac{dM_1}{dt} = D_2\frac{C_2(L_1,t) - C_2(L_1,t)}{L_2 - L_1} = D_2\frac{K_{12}C_1(L_1,t) - C_3(L_2,t)/K_{23}}{L_2 - L_1}$$

$$M_t = C_3(L_2,t)V = \frac{D_2(K_{12}C_1(L_1,t) - C_3(L_2,t)/K_{23})A\,t}{L_2 - L_1}$$

4b. Diffusion Mode

$$C_1 = \frac{C_3}{K_{12}K_{23}} + \left(C_0 - \frac{C_3}{K_{12}K_{23}}\right)\exp\left[\frac{-D_2 t}{L_1(L_2 - L_1)}\right]$$

$$C_2 = \frac{C_3/K_{23} - K_{12}C_1}{L_2 - L_1}x - \frac{L_1(C_3/K_{23} - K_{12}C_1)}{L_2 - L_1} + K_{12}C_1$$

$D_2 \ll D_1$ and $D_2 \ll D_3$ (more than about 10 times)

$$D_1\frac{\partial C_1}{\partial x} = D_2\frac{\partial C_2}{\partial x}\bigg|_{x=L_1} \quad \text{and} \quad C_2(L_1,t) = K_{12}C_1(L_1,t) \quad \text{at } t > 0$$

$$D_2\frac{\partial C_2}{\partial x} = D_3\frac{\partial C_3}{\partial x}\bigg|_{x=L_2} \quad \text{and} \quad C_3(L_2,t) = K_{23}C_2(L_2,t) \quad \text{at } t > 0$$

Figure 6 Equations and conditions for mass transfer models.

Nomenclature for Figure 6

A	Surface area of the matrix
C_0	Initial concentration of a migrant in layer $0 \le x \le L_1$
C_1	Concentrations of a migrant in layer $0 \le x \le L_1$
C_2	Concentrations of a migrant in layer $L_1 \le x \le L_2$
C_3	Concentrations of a migrant in layer $L_2 \le x \le L_3$
D_1	Diffusivities of a migrant in layer $0 < x < L_1$
D_2	Diffusivities of a migrant in layer $L_1 \le x \le L_2$
D_3	Diffusivities of a migrant in layer $x > L_2$
J	Mass transfer flux
K_{12}	$= C_2/C_1$ at $x = L_1$
K_{23}	$= C_3/C_2$ at $x = L_2$
h	$= (1 - m)/(1 - m)$
L	Half thickness of the sheet
L_1	Thickness of layer 1
L_2	Thickness of layer 2
L_3	Thickness of fluid layer 3 ($= V/A$)
m	$= (D_1/D_2)^{1/2}$
M_t	Amount of released migrant at time t
M_∞	Amount of released migrant at infinite time
t	Time
V	Volume of immersing fluid (layer 3)
x	Distance from the origin

Equation 1 is also very useful for determining the diffusivity D_1 of an active substance in the packaging material by use of a migration cell and water (or a liquid food directly). The thin packaging material is immersed in water or liquid food, and the concentration increase of migrant in the liquid is observed periodically to calculate the fractional migration with time. Because Equation 1 has only one independent variable, time t, it is very simple to estimate D_1 with given thickness L and measured fractional migration M_t/M_∞. However, it cannot explain the concentration profile at any position in the packaging material.

The other approach is to use a two-layer diffusion model [Figure 5(a)]. The model (Equation 2) used by Luicov (1968) and Han (1996) is one of the two-layer diffusion models used to explain concentration distributions in the two layers. Luicov (1968) built up heat-transfer models with different heat conductivities for the double-layer material, and Han (1996) modified Luicov's model for the two-layer mass-transfer model. Equation 2 has the position variable x in the mass-transfer direction, unlike the migration model. Antimicrobial agents are incorporated into the first packaging layer ($0 < x < L$) and diffuse into the second food layer ($L < x$) Concentrations of moving substance in the packaging layer (C_1 and food layer C_2) are related to the diffusivity ratio m as well as their diffusivities in each layer (D_1 and D_2). Therefore, the diffusivity in one layer affects the diffusion profile in the other layer.

Han (1996) also developed Equation 3 from his model (Equation 2) to contain the partition coefficient $K = C_2/C_1$. Figure 7 shows the computer simulation result of Equation 3.

A three-layer model (outer package/inner package/food) is also usable [Figure 5(b)] if the packaging system requires a control layer as an inner packaging layer to control mass transfer accurately. The control layer (inner layer) usually has lower diffusivity than the matrix (outer) layer and the food layer. Therefore, the mass transfer is highly dependent on the permeation of the active substance through the control layer rather than the diffusion in the outer layer or food layer.

Equations 4a and b in Figure 6 are the mass transfer models for the three-layer system, in which the control layer can control the release rate and the

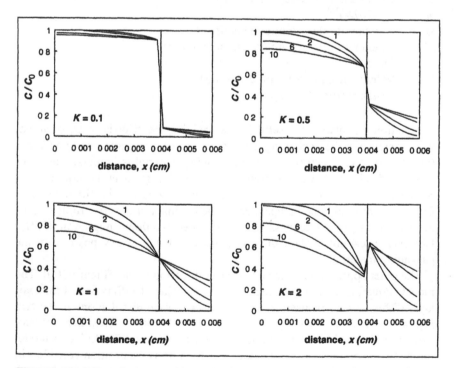

Figure 7 Simulations of release profiles with different partition coefficients K, at L_1. $D_1 = D_2 = 1 \times 10^{-8}$ cm^2/sec, $L_1 = 40$ mm, and numbers on lines are times $t = 1, 2, 6,$ and 10 min.

overall release profile. These models need experimental data for the concentration C_3 at the finite volume of layer 3 because they do not have a mathematical solution for C_3. However, the models can be used for the controlled-release membrane system and transdermal system for drug delivery, as well as for food packaging. Diffusivity of a migrant in the control layer (D_2) can be easily calculated by measuring M_t with time using Equation 4a. During the steady-state period, D_2 can be obtained from the slope of a plot of M_1 versus time. The rate of reduction of C_1 in layer 1 is equal to the flux J in layer 2. Therefore, the equation for C_1 was obtained mathematically, and the linear equation for C_2 in layer 2 was obtained by the steady-state assumption (Equation 4b).

When C_3, the concentration of the migrant in fluid layer 3 (which may be water or liquid food), is measured experimentally, the distribution of C_1 and C_2 can be obtained from Equation 4b. Equations 4a and b work only when the diffusivity is lowest in the control layer compared to the matrix and food layers (i.e., when $D_2 << D_1$ and D_3). They also assume steady-state mass transfer in the control layer. Therefore, at the very early stage of mass transfer before reaching steady state at the control layer, these models are not good at explaining the concentration profile in the control layer. These conditions can usually be ignored in any case of packaging system, however, because the high barrier properties of packaging materials satisfy the D_2 = lowest condition, and the storage/distribution period of packaged products is long enough to reach steady state.

Mass Transfer in Laminated Films

In laminated multilayer films or containers, thin films can provide specific release rates with assumptions of (pseudo-) steady-state condition when they are used as a control layer. Because the antimicrobial packaging systems will use chemical (or naturally existing) antimicrobial agents, the overall mass transfer will follow the solute diffusion and permeation rules, unlike the gas transfer case. The solute chemical transfer is caused by the concentration difference ΔC (in mole/L or %). The concentration difference of the solute chemical can be converted to the difference in partial pressure Δp (in atm) because the moving substance is not in the gaseous phase.

In the case of gas transfer, the permeability P (cm^2/sec·atm) can be expressed by the diffusivity D (cm^2/sec) and the solubility S (atm^{-1}). Therefore, $P\Delta p/L$, which is the flux of the gas transfer, can be converted to $D\Delta C/L$. However, in the case of solute transfer, the permeation has the driving force of ΔC and does not contain the partial pressure. The units of the permeability P of the solute (cm/sec) differ from those of the permeability of the gas (cm^2/sec·atm). The $P\Delta C$ of the solute transfer, which is the flux, can be converted to $DK\Delta C/L$.

Therefore, the overall mass transfer resistance in multilayer structure can be expressed as follows (Fan and Singh, 1989; Flynn et al., 1974):

$$R_T = \frac{1}{P_T} = \sum_{i=1}^{n} R_i = \sum_{i=1}^{n} \frac{1}{P_i} = \frac{L_1}{D_1 K_1} + \frac{L_2}{D_2 K_2} + \ldots + \frac{L_n}{D_n K_n}$$

where R and P are the overall resistance and permeability of the laminated material; D_i and L_i are the diffusivity and thickness of the ith layer of individual materials, respectively; and K_i is the partition coefficient of the diffusing substance at the interface between the ith and $(i - 1)$th layers.

When partition coefficients are ignored (i.e., given a value of 1), the above equation is simplified to:

$$R_T = \frac{1}{P_T} = \frac{L_T}{D_T}$$

where D_T and L_T are the overall diffusivity and total thickness, respectively, of the laminated film.

The lag time t_L of diffusion, which is the time taken for the diffusant to penetrate through the film material, was defined as $t_L = L^2/6D$. Flynn et al. (1972, 1974) calculated the lag time through multilayer materials using the following equation:

$$T_L = \frac{\left(\sum L_i \right)^2}{6D_T}$$

This lag time is an initial delay of active substance penetration through the control layer and/or inner laminated layers. It does not show the antimicrobial activity during the early stage just after lamination. It is the very early period of mass transfer that does not follow the steady-state assumption of film penetration. The diffusant in the reservoir matrix layer starts to move into the neighboring control layer (and/or other inner layers) as soon as the lamination process is completed. During film storage and the processes of printing, slitting, and shipping to the food industry, the diffusant in the matrix layer moves into the neighboring layer until the mass transfer is equilibrated. Thus, the lag period of a composite film is affected by the previous converting process time and storage period before food products are packaged. If the packaged food is contaminated and microbial growth occurs during this lag period, the antimicrobial packaging system cannot control the spoilage. However, package materials usually stay in the storage room after being supplied to the food packager. If the date supplied materials are received is recorded, spoilage during the lag period could be prevented.

The overall release profile in the multilayer structure is affected by the slowest diffusion layer (the control layer), which has the smallest diffusivity, such as the bottlenecking of sequential transportation processes. Keeping the high diffusivity of the active substance in the reservoir matrix (D_1) constant, the overall release profile can be controlled by the diffusivity of the neighboring control layer (D_2) slowly and accurately.

VERIFICATION OF ANTIMICROBIAL ACTIVITY

The antimicrobial activity of the package materials could be measured by microbiology experiments. Before food samples are packaged in the antimicrobial package materials, target microorganisms may be inoculated onto the surface of the food or mixed into the food samples. Measuring growth by counting the microorganisms with incubation time can provide the characteristic values of growth rate at the exponential growth phase, maximum growth at the stationary phase, and initial lag period. These values can be compared to conventional packaging or reference samples without packaging.

Reduced growth rate and reduced maximum growth population could indicate improved microbial safety, and the extended lag period would show the prolonged shelf life with microbial quality assurance. The growth rate can be easily obtained from the slope of the plot of logarithmic transformed growth versus time, using the growth data from the exponential growth period. Linear regression analysis will provide the slope value as a parameter estimate of the time variable (Han and Floros, 1997, 1998a). The lag period can also be obtained by the same regression analysis. Extrapolation of the linear line to the *x*-intercept will show the lag period. Maximum growth population can be obtained from the converging growth value or a peak value of growth at the stationary phase of the growth curve directly.

SORBATE-RELEASING LOW DENSITY POLYETHYLENE FILM

Sorbate-releasing plastic film for cheese packaging is a good example of successful research and development of antimicrobial packaging. To develop the film, the diffusivities of the food preservative potassium sorbate through various plastic films and cheeses were first determined as described by Han and Floros (1998b, 1998c). Plastics have diffusivities ranging from 1×10^{-8} through 1×10^{-14} cm^2/sec for potassium sorbate, while cheeses have diffusivities ranging from 1×10^{-6} through 1×10^{-7} cm^2/sec at 25°C. Cheeses and solid foods usually have diffusivities that are 10^2 to 10^6 times larger than those of plastics for potassium sorbate. High-moisture foods and liquid foods may have much higher diffusivities than dry and solid foods.

Because diffusivity in plastics is smaller than that in foods, release of most antimicrobial agents, including potassium sorbate, highly depends on the lowest diffusivities of the active agents in the packaging materials. Using the two-layer mass transfer model (Equation 2), the release profiles of potassium sorbate from plastic films (low density polyethylene and polypropylene) into processed American cheese and mozzarella cheese were simulated by computer programming (Han, 1996). Figure 8 shows the simulation results with given diffusivities of potassium sorbate in low density polyethylene, polypropylene, processed American cheese, and mozzarella cheese.

Because cheese has surface microbial (usually fungal) contamination due to post-process contamination, the potassium sorbate concentration on the cheese surface is critical and must be maintained above 0.1% to inhibit mold growth. When processed American cheese was packaged with 40-μm-thick high density polyethylene (HDPE) films containing 10% sorbate initially and stored at room temperature, the surface concentration was maintained above 0.1% for about five months after packaging (Han, 1996). Therefore, the cheese packaging enables the cheese to be microbe-free for five months at room temperature.

A suggested antimicrobial film casting process is shown in Figure 9, which uses the low density polyethylene as a matrix layer for the antimicrobial agent. The plastic resin and the antimicrobial agents are mixed, extruded, and pelletized to produce masterbatch resins. The pellets are then added to clear plastic resin to provide the expected amount of the masterbatch to produce the final

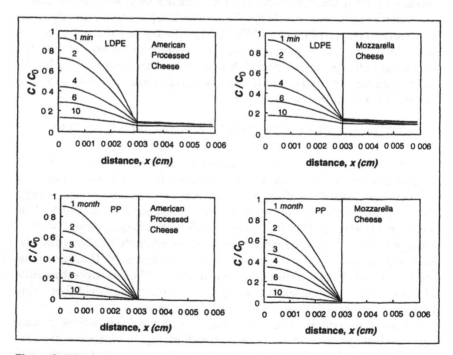

Figure 8 Release profile of potassium sorbate from plastic films into cheeses (film thickness $L_1 = 30$ μm).

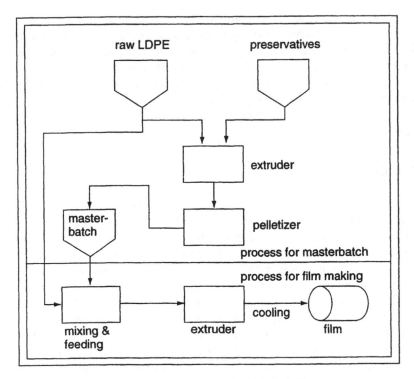

Figure 9 Suggested process for producing antimicrobial plastic film

matrix film or sheet incorporating the proper amount of the antimicrobial agent. This process would have a homogeneous concentration of the antimicrobial agent in the final product because of the double extrusion. The process is expected to produce a uniform concentration distribution of the active substance in the final package material and a uniform controlled release rate.

ANTIMICROBIAL EDIBLE COATINGS AND FILMS

Edible coatings and films have a variety of advantages such as biodegradability, edibility, biocompatibility, aesthetic appearance, and barrier properties against oxygen and physical stress. They can also serve as a carrier for edible antimicrobial agents and preservatives. For example, whey protein coatings and films can incorporate adequate amounts of edible antimicrobial agents (e.g., lysozyme, nisin, potassium sorbate, EDTA), as well as a plasticizer (e.g., glycerin or sorbitol). Because the charge density and the cavity size of the three-dimensional protein network of the whey protein film can be adjusted by altering the pH and the volume ratio of the whey protein and the plasticizer, the diffusion rate of the incorporated antimicrobial agents can be controlled.

Lysozyme-impregnated whey protein film is effective against a spoilage microorganism, *Brochothrix thermosphacta* The lysozyme, slowly released from the film, effectively inhibited the growth of the microorganism, as indicated by the clear inhibitory zone around the film discs. The lysozyme-impregnated whey protein film showed great clarity and maintained the tensile strength at concentrations of up to 100 mg of lysozyme/g of dried film. This film may have great potential as a microbial hurdle against gram-positive spoilage and pathogenic bacteria.

CONCLUSIONS

Antimicrobial packaging is a promising form of active food packaging. Although many packaged perishable food products are heat sterilized or have a self-protecting immune system, microbial contamination could occur on the surface or damaged area of the food through package defects or restorage after opening. The antimicrobial substances incorporated into package materials can control microbial contamination by reducing the growth rate and maximum growth population and extending the lag period of the target microorganism.

Package materials incorporating antimicrobial agents have been a packaging development objective for many years, with relatively few commercial successes, except in Japan. In Japan, for reasons that are not clear, compounds such as silver salts on zeolite, allyisothiocyanate extracted from horseradish or mustard, and hinokotiol have been accepted by regulatory authorities as both safe and efficacious. The reported effectiveness of these materials suggests further investigation to confirm the microbicidal activity and potential side effects in food package materials. A challenge is a persistent odor.

Silver salt-containing package materials appear to function on contact, and so are suitable mainly for surface microorganisms.

The other materials appear to possess some microbicidal properties through volatilization. Some Japanese suppliers claim their AIT compounds do not impart a flavor, although all reported U.S. testing indicates strong pungent odors.

Research on chlorine dioxide as a broad-spectrum antimicrobial, which functions by volatilization at a distance from its source, indicates that this compound could be considered as a serious contender in the area of active packaging antimicrobials. As with so many other such materials, regulatory acceptance is a hurdle.

Ethanol's effectiveness as an antimicrobial, particularly for intermediate-moisture products such as soft bakery goods and cheese, has been demonstrated both in the laboratory and in commercial practice. Concerns about its flavor have been, in part, alleviated by incorporation of counter flavors that may also enhance the basic product flavors. Concerns about ethanol because of consumer activism represent a nontechnical challenge of great importance.

BIBLIOGRAPHY

Anonymous 1978. "Antibacterial and antifungal composition and web material coated therein." British Patent 1,518,650.

Anonymous. 1979. "Agent for influencing bacterial attack by materials coming into contact with the human body, preferably food packaging materials, and process for its use." German Patent 2,749,528. May 3.

Anonymous 1991a. "Antimicrobial Zeomic®" Shinagawa Fuel Co., Ltd.

Anonymous. 1991b. "Zeomic." Mitsubishi Corporation.

Anonymous. 1997. "New Antibiotic Pouch for Long-Term Drinking Water Preservation" *Packaging Trends Japan,* 97:2.

Appendini, P. and J. H. Hotchkiss. 1996. "Immobilization of Lysozyme on Synthetic Polymers for the Application to Food Packages." Presented at the annual meeting of Institute of Food Technologists, New Orleans, Louisiana, June 22–26.

Austin, P. R., C. J. Brine, J. E. Castle, and Zikakis. 1981. "Chitin: New Facets of Research." *Science,* 212:749

Baron, J. K and S. S. Sumner. 1993. "Antimicrobial Containing Edible Films as an Inhibitory System to Control Microbial Growth on Meat Products" *Journal of Food Protection,* 56:916.

Ben-Yehoshua, S., B. Y. Shapiro, Y Gutter, and E Barak. 1987. "Comparative Effects of Applying Imazalil by Dipping or by Incorporation into the Plastic Film on Decay Control, Distribution and Persistence of This Fungicide in Shamouti Oranges Individually Seal-Packaged." *J. Plastic Film and Sheeting,* 3(1):9–22. January.

Bomar, M. 1968. "Pesticidal Properties of Organotin Compounds with Regard to the Utilization in Packaging Technology." *Obaly,* 14(4):107–109

Brody, A. L. and J A. Budny. 1995. "Enzymes as Active Packaging Agents." In: M. L. Rooney (ed.), *Active Food Packaging.* Glasgow, UK, Blackie Academic & Professional. pp. 174–192.

Cardarelli, N. 1976. *Controlled Release Pesticides Formulations.* Cleveland, Ohio, CRC Press, Inc. pp 120–124.

Carslaw, H S. and J. C. Jaeger. 1947. *Conduction of Heat in Solids.* Oxford, England: Clarendon Press.

Carslaw, H. S. and J C. Jaeger. 1959. *Conduction of Heat in Solids,* 2nd Ed. Oxford, England, Clarendon Press.

Chen, M.-C., G. H C. Yeh, and B.-H Chiang. 1996. "Antimicrobial and Physicochemical Properties of Methyl Cellulose and Chitosan Films Containing a Preservative." *Journal of Food Processing & Preservation,* 20:379–390.

Crank, J. 1975 *The Mathematics of Diffusion,* 2nd Ed. Oxford, England, Clarendon Press.

Cuera, R. G., G. Osuji, and A. Washington. 1991. "N-Carboxymethyl Chitosan Inhibition of Aflatoxin Production—Role of Zinc" *Biotech. Lett.,* 13:441–444.

Daeschel, M. A., J. McGuire, and H. Al-Makhlafi. 1992. "Antimicrobial Activity of Nisin Absorbed to Hydrophilic and Hydrophobic Silicon Surfaces." *Journal of Food Protection,* 55:731–735.

Fan, L. T. and S. K Singh. 1989. *Controlled Release, A Quantitative Treatment,* New York: Springer-Verlag. pp. 9–88.

Feld, C. E., L. F Pivarnick, S. M. Barnett, and A. Rand. 1986. "Utilization of Glucose Oxidase for Extending the Shelflife of Fish." *Journal of Food Science,* 51:66–70.

Floros, J. D., L. L. Dock, and J. H. Han. 1997. "Active Packaging Technologies and Applications." *Food, Cosmetic & Drug Packaging,* 20(1):10–17.

Flynn, G. L., O. S. Carpenter, and S. H. Valkowsky. 1972. "Total Mathematical Resolution of Diffusion Layer; Control of Barrier Flux." *Journal of Pharmaceutical Science,* 61:312–314.

Flynn, G., S. Valkowsky, and T. J. Roseman. 1974. "Mass Transport Phenomena and Models: Theoretical Concepts." *Journal of Pharmaceutical Science,* 63:479–510.

Ghosh, K. G., A N Srivatsava, N. Nirmala, and T. R. Sharma. 1973. "Development and Application of Fungistatic Wrappers in Food Preservation. Part I. Wrappers Obtained by Impregnation Method." *Journal of Food Science & Technology,* 10(4):105–110.

Ghosh, K. G., A. N. Srivatsava, N. Nirmala, and T. R. Sharma. 1977. "Development and Application of Fungistatic Wrappers in Food Preservation. Part II. Wrappers Made by Coating Process." *Journal of Food Science & Technology,* 14(6): 261–264.

Gontard, N. 1997. "Active Packaging." *Proceedings of Workshop sobre Biopolimeros,* (April 22–24, 1997. Univ. de Sao Paulo), ed. P. J. do A. Sobral and G. Chuzel, pp. 23–27. Prassununga, FZEA, Brazil.

Gray, Peter. 2000. "Generation of Active Microatmosphere™ Environments from and in Packages." *Proceedings, International Conference on Active and Intelligent Packaging.* Campden & Chorleywood Food Research Association, U.K. September, 2000.

Hagelstein, A. E., D. G. Hoover, J. S. Paik, and M. H. Kelley. 1995. "Potential of Antimicrobial Nylon as a Food Package." Presented at the annual meeting of Institute of Food Technologists, Anaheim, California, June 3–7.

Halek, G. W. and A. Garg. 1989. "Fungal Inhibition by a Fungicide Coupled to an Ionomeric Film." *Journal of Food Safety,* 9:215–222.

Han, J. H. 1996. "Modeling the Inhibition Kinetics and the Mass Transfer of Controlled Releasing Potassium Sorbate to Develop an Antimicrobial Polymer for Food Packaging." Ph.D dissertation, Purdue University, West Lafayette, Indiana.

Han, Jung H. 2000. "Antimicrobial Food Packaging." *Food Technology,* 54(3):56–65. March.

Han, J. H. and J. D Floros. 1996. "The Effect of Temperature and Concentration on the Diffusivity of Potassium Sorbate through Plastic Films." The annual meeting of Institute of Food Technologists, New Orleans, Louisiana, June 22–26.

Han, J. H. and J. D. Floros. 1997. "Casting Antimicrobial Packaging Films and Measuring Their Physical Properties and Antimicrobial Activity." *Journal of Plastic Film and Sheeting,* 13:287–298.

Han, J. H. and J. D. Floros. 1998a. "Modeling the Growth Inhibition Kinetics of Baker's Yeast by Potassium Sorbate using Statistical Approaches." *Journal of Food Science,* 63:12–14.

Han, J. H. and J. D. Floros. 1998b. "Potassium Sorbate Diffusivity in American Processed and Mozzarella Cheeses " *Journal of Food Science,* 63:435–437.

Han, J. H. and J. D. Floros. 1998c. "Simulating Diffusion Model and Determining Diffusivity of Potassium Sorbate Through Plastics to Develop Antimicrobial Packaging Film." *Journal of Food Processing & Preservation,* 22(2):107–122.

Han, J. H. and J. D. Floros. 1999. "Modeling Antimicrobial Activity Loss of Potassium Sorbate Against Baker's Yeast after Heat Process to Develop Antimicrobial Food Packaging Materials." *Food Science Biotechnology,* 8(1):11–14.

Hansen, R. E., C. G. Rippl, D. G. Midkiff, and J. G. Neuwirth 1989 "Antimicrobial absorbent food pad." U.S Patent 4,865,855.

Hoojjat, P., B. Honte, R. Hernandez, J. Giacin, and J. Miltz. 1987. "Mass Transfer of BHT from HDPE Film and Its Influence on Product Stability." *Journal of Packaging Technology,* 1(3):78.

Hotchkiss, Joseph W. 1995. "Influence of New Packaging Technologies on the Microbial Safety of Muscle Foods." The annual meeting of Institute of Food Technologists, Anaheim, California, June 3–7.

Hotchkiss, J. H. 1997. "Food-Packaging Interactions Influencing Quality and Safety." *Food Additives Contaminants,* 14:601–607.

Huang, L. J., C. H. Huang, and Y.-M. Weng. 1997. "Using Antimicrobial Polyethylene Films and Minimal Microwave Heating to Control the Microbial Growth of Tilapia Fillets During Cold Storage." *Food Science Taiwan,* 24:263–268. Cited in *Food Science & Technology Abstracts.*

Hurst, A. and D. G. Hoover. 1993. *Antimicrobials in Foods.* P. Michael Davidson and Alfred Larry Branen (eds.). New York: Marcel Dekker, Inc.

Intili, H. S. 1985. "Antimicrobial paper." U. S. Patent 4,533,435. August 6.

Ioka, S., Y. Kataoka, and H. Yagi 1975. "Fungi-resistant paper." Japanese Patent Kokai 105,906/75. August 21.

Ishitani, T. 1995. "Active Packaging for Food Quality Preservation in Japan." In: P. Ackermann, M. Jagerstad, and T. Ohlsson (eds.). *Food and Food Packaging Materials—Chemical Interactions,* Cambridge, England: Royal Society of Chemistry. pp. 177–188. Cited in Hotchkiss (1997).

Kanie, T. and K. Ichikawa. 1991 "Antimicrobial paper." Japanese Patent Kokai 898/91. January 7.

Kato, J., H. Komiya, and Y. Hata 1983. "Antimicrobial sheet." Japanese Patent Kokai 76,599/83. May 9.

Khadzzhieva, L. and Ts. Kristov. 1975. "Fungistatic Properties of Some Reagents Used for Treating Fruit Packaging Papers." *Tseluloza Khartiya,* 6(2):17–22.

Knorr, D. 1991. "Recovery and Utilization of Chitin and Chitosan in Food Processing Waste Management." Metals adsorption table. *Food Technology,* 45(1):114–122.

Krochta, J. M. and C De Mulder-Johnston. 1997 "Edible and Biodegradable Polymer Films: Challenges and Opportunities." *Food Technology,* 51(2):61–74.

Labuza, T. P. 1990. "Active Food Packaging Technologies." *Food Science & Technology Today,* 4(1):53–56

Labuza, T. P. and W. M. Breene. 1988. "Applications of 'Active Packaging' for Improvement of Shelf-Life and Nutritional Quality of Fresh and Extended Shelf-Life Foods." *Journal of Food Processing & Preservation,* 13:1–69.

Lee, D. S., Y. L. Hwang, and S. H. Cho. 1998 "Developing Antimicrobial Packaging Film for Curled Lettuce and Soybean Sprouts." *Food Science Biotechnology,* 7(2):117–121.

Luicov, A. V. 1968. *Analytical Heat Diffusion Theory,* J. P. Harnett (ed.). New York: Academic Press.

Mayer, J. M and D. Kaplan. 1991 "Method of forming a cross-linked chitosan polymer and product thereof" U.S. Patent 5,015,293. May 14.

Mayer, J. M., B. Wiley, K. Henderson, and D. Kaplan. 1989. "Physical Properties of Films Produced from the Biopolymers Pullulan and Chitosan Produced by *Aureobasidicum pullulans* and *Mucor rouxii*: Oxygen Permeability, Tensile Strength, etc." Abstracts, Annual Meeting. American Society of Microbiologists.

Mayer, J. M., M. Greenberger, and D. L. Kaplan. 1994. "Biopolymers for Packaging/ Coatings and Standard Methods to Quantify Rates of Degradation." In K. L. Garg, N. Garg, and K. G. Mukerji (eds.). *Recent Advances in Biodeterioration and Biodegradation, II*, pp. 41–70. Calcutta: Naya Prokash.

Melnick, D. and F. H. Luckmann. 1954a. "Sorbic Acid as a Fungistatic Agent for Foods. III. Spectrophotometric Determination of Sorbic Acid in Cheese and in Cheese Wrappers." *Food Research,* 19:20–27.

Melnick, D. and F. H. Luckmann. 1954b. "Sorbic Acid as a Fungistatic Agent for Foods. IV. Migration of Sorbic Acid from Wrapper into Cheese." *Food Research,* 19:28–32.

Melnick, D., F. H. Luckmann, C. M. Gooding. 1954. "Sorbic Acid as a Fungistatic Agent for Foods. VI. Metabolic Degradation of Sorbic Acid in Cheese by Molds and the Mechanism of Mold Inhibition." *Food Research,* 19:44–57.

Miller, W. R., D. H. Spalding, L. A. Risse, and V. Chew. 1984. "The Effects of an Imazalil-Impregnated Film with Chlorine and Imazalil to Control Decay of Bell Peppers." *Proceedings of Florida State Horticultural Society,* 97:108–111.

Ming, X., G. H. Weber, J. W. Ayers, and W. E. Sandine. 1997. "Bacteriocins Applied to Food Packaging Materials to Inhibit *Listeria monocytogenes* on Meats." *Journal of Food Science,* 62(2):413–415.

Mori, Y., K. Okunishi, Y. Kataoka, and E. Adachi. 1987. "Microbicide-impregnated paper bags for fruit protection." Japanese Patent Kokai 39,505/87. February 20.

Nicholson, Myron D 1997. "The Role of Natural Antimicrobials in Food/Packaging Preservation." *Proceedings of Future Pack '97.* George O. Schroeder Associates, Inc., Appleton, Wisconsin.

Ouattara, B., S. Simard, G. J. P. Piette, R. A Holley, and A. Begin. 1999. "Diffusion of Acetic and Propionic Acids from Chitosan Films Immersed in Water." Presented at the annual meeting of Institute of Food Technologists, Chicago, Illinois, July 24–28.

Padgett, T., I. Y. Han, and P L. Dawson. 1998. "Incorporation of Food-Grade Antimicrobial Compounds into Biodegradable Packaging Films." *Journal of Food Protection,* 61:1330–1335.

Padmanabha Reddy, V. and M. Mohamed Habibullah Khan. 1993. "Effect of Antimicrobial Agents and Packaging Materials on the Microbial Quality of Khoa." *Journal of Food Science and Technology* (India), 30(2):130–131.

Paik, J. S. and M. J. Kelley. 1995. "Photoprocessing Method of Imparting Antimicrobial Activity to Packaging Film." Presented at the annual meeting of Institute of Food Technologists, Chicago, Illinois, July 24–28.

Pennisi, E. 1992. "Sealed in Edible Film." *Science News,* 141:12–13.

Popper, L. and D. Knorr. 1990. "Applications of High-Pressure Homogenization for Food Preservation." *Food Technology,* 44(7):84–89.

Porter, W. L. 1993. "Paradoxical Behavior of Antioxidants in Food and Biological Systems." *Toxicol. Ind. Health,* 9:93–122.

Porter, William L., Karen R. Conca, R. Victor Lachica, Jean M. Mayer, and E. Ray Pariser. 1995. "Chitin and Chitosan as Navel Protective Food Ingredients." Research and Development Associates Annual Meeting, 1995 Vol. 48, No. 1.

Ralston, G. B., M. V. Tracey, and P. M Wrench. 1964. The Inhibition of Fermentation in Baker's Yeast by Chitosan." *Biochem. Biophys. Acta,* 93:652.

Rice, J. 1995 "Antimicrobial Polymer Food Packaging." *Food Processing,* 56(4):56, 58.

Rico-Pena, D. C. and J. A. Torres. 1991. "Sorbic Acid and Potassium Sorbate Permeability of an Edible Methylcellulose-Palmitic Acid Film: Water Activity and pH Effects." *Journal of Food Science,* 56·497–499.

Rooney, M. L. 1981. Oxygen Scavenging from Air in Package Headspace by Singlet Oxygen Reactions in Polymer Media " *Journal of Food Science,* 47:291–294, 298.

Rooney, M. L. 1996. Personal communication. CSIRO, Australia

Rubinstein, W. S 2000. "Microban® Antibacterial Protection for the Food Industry." *Proceedings, International Conference on Active and Intelligent Packaging.* Campden & Chorleywood Food Research Association, U.K. September, 2000

Sacharow, S. 1988. "Freshness Enhancers: The Control in Controlled Atmosphere Packaging." *Preserved Foods,* 157(5):121–122.

Seabrook, G. and R. C. Heymann. 1996. "Compositions containing antimicrobial agents and methods for making and using same." U S. Patent 5,554,373, September 10.

Shapero, M., D. Nelson, and T P Labuza. 1978. "Ethanol Inhibition of *Staphylococcus aureus* at Limited Water Activity." *Journal of Food Science,* 43:1467–1469.

Shetty, K. K. and R. B. Dwelle. 1990. "Disease and Sprout Control in Individually Film Wrapped Potatoes." *American Potato Journal,* 67(10):705–718.

Smith, D. P. and N. J. Rollin. 1954a. "Sorbic Acid as a Fungistatic Agent for Foods. VII. Effectiveness of Sorbic Acid in Protecting Cheese." *Food Research,* 19:50–65.

Smith, D. P. and N. J. Rollin. 1954b. "Sorbic Acid as a Fungistatic Agent for Foods. VIII Need and Efficacy in Protecting Packaged Cheese." *Food Technology,* 8(3):133–135

Smith, J. P., B. Ooraikul, W. J. Koersen, E. D. Jackson, and R. A. Lawrence 1986. "Novel Approach to Oxygen Control on Modified Atmosphere Packaging of Bakery Products." *Food Microbiology,* 3:315–320.

Smith, J. P., B. Ooraikul, W. J. Koersen, F. R. van de Voort, E D Jackson, and R. A. Lawrence 1987 "Shelf Life Extension of a Bakery Product using Ethanol Vapor." *Food Microbiology,* 4:329–337.

Smith, J P. 1988. "Shelf Life Extension of a Fruit Filled Bakery Product Using Ethanol Vapor" *CAP '88 International Conference on Controlled/Modified Atmosphere/ Vacuum Packaging, 1988 Conference Proceedings.* Schotland Business Research, Inc., Princeton, New Jersey.

Smith, J. P., F. R. Van de Voort, and A. Lambert. 1989. "Food and Its Relation to Interactive Packaging." *Canadian Institute of Food Science Technology,* 22(4):327–330, October.

Smith, J. P., H. A. Ramaswamy, and B. K. Simpson. 1990. "Developments in Food Packaging Technology. Part II: Storage Aspects." *Trends Food Science Technology,* pp. 111–118, November.

Smith, J. P, J. Hoshino, and Y. Abe. 1995 "Interactive Packaging Involving Sachet Technology." In: Michael L. Rooney (ed.). *Active Packaging,* Glasgow, UK: Blackie Academic & Professional.

Stilles, M E 1996 "Biopreservation by Lactic Acid Bacteria." *Antonie van Leeuwenhoek,* 70:331–345. Netherlands: Kluwer Academic Publishers.

Suzuki, K., K. Hokita, and T. Itoh. 1991. "Tectosilicate antimicrobial composition." Japanese Patent Kokai 120,204/91 May 22.

Takita, K., H. Sugiyama, and Y. Konagai. 1975. "Mold-resistant packaging paper." Japanese Patent Kokai 88,311/75. July 16.

Thonack, William G. 1988. "Ethicap—Introduction" *CAP '88 International Conference on Controlled/Modified Atmosphere/Vacuum Packaging, 1988 Proceedings,* Schotland Business Research, Inc., Princeton, New Jersey.

Torres, J. A., J. O. Bouzas, and M. Karel. 1985. "Microbial Stabilization of Intermediate Moisture Food Surfaces. III. Effectiveness of Surface Preservative Concentration and Surface pH Control on Microbial Stability of an Intermediate Moisture Cheese Analog." *Journal of Food Processing & Preservation,* 9:107–119.

U.S. Department of Agriculture (Assignee). 1994. "Inhibition of post harvest fruit decay by 2-nonanone." U.S. Patent 5,334,619. August 2.

Vermeiren, Lieve. 2000 "Potential Applications of Antimicrobial Films for Food Packaging." *Proceedings, International Conference on Active and Intelligent Packaging.* Campden & Chorleywood Food Research Association, U.K. September.

Vojdani, F. and J. A. Torres. 1990 "Potassium Sorbate Permeability of Methylcellulose and Hydroxypropyl Methylcellulose Coatings: Effect of Fatty Acid." *Journal of Food Science,* 55:841–846.

Wellinghoff, Stephen T. 1995. "Keeping Food Fresh Longer." *Technology Today.* June.

Weng, Y.-M. and M.-J. Chen. 1997. "Sorbic Anhydride as Antimycotic Additive in Polyethylene Food Packaging Films." *Lebensm, Wiss. U. Technol.,* 30:485–487.

Weng, Y.-M. and J H. Hotchkiss. 1992. "Inhibition of Surface Molds on Cheese by Polyethylene Film Containing the Antimycotic Imazalil." *Journal of Food Protection,* 55:367–369.

Weng, Y.-M. and J. H. Hotchkiss. 1993. "Anhydrides as Antimycotic Agents Added to Polyethylene Films for Food Packaging." *Packaging Technology & Science,* 6(3):123–128. May–June.

Weng, Y.-M., M.-J. Chen, and W. L. Chen. 1997. "Antimicrobial Food Packaging Materials from Poly(ethylene-comethacrylic acid)." Presented at the annual meeting of Institute of Food Technologists, Orlando, Florida, June 14–18.

Wilhoit, D. L. 1996. "Surface treatment of foodstuffs with antimicrobial compositions." U.S. Patent 5,573,801. November 12.

Yoshida, H., K. Matsuo, and J. Yagi. 1993. "Antimicrobial paper." Japanese Patent Kokai 93,397/93. April 16.

Yoshioka, Y., K. Takahashi, and K. Hashida. 1974. "Paper having antiseptic nature." Japanese Patent 23,245/74. June 14.

Zimina, I. A., M. A. Solyanik, and N. D. Mikhnovaskaya. 1975. "Methods of Evaluating the Quality of Fungicidal Paper." *Sb. Tr. Ukr. Nauch.-Issled. Inst. Tsellyul.-Bumazh Prom.,* 18:122–125.

Zimina, I. A., Z. L. Zosim, and L. A. Pirogova. 1974. "Fungicidal Activity of Packaging Food Paper." *Sb. Tr. Ukr. NII Tsellyul.-Bumazh. Prom.,* 16:122–125.

CHAPTER 11

Other Systems

CARBON DIOXIDE ABSORBERS

ALTHOUGH carbon dioxide often exerts a microbiological inhibitory effect in modified atmosphere packaging, excess carbon dioxide may at times adversely affect the product or counter the inhibitory effect. Therefore, some food-preservation packaging systems are engineered to remove carbon dioxide.

In their oxygen-scavenging functions, iron oxide oxygen scavengers also absorb carbon dioxide, although the absorption is preferential for oxygen over carbon dioxide. Mitsubishi Gas Chemical Company offers (and uses in roasted and ground coffee packaging) sachets specifically engineered for carbon dioxide absorption. Previously, General Foods (now Kraft) Maxwell House brand employed iron oxide sachets within steel cans of roasted and ground coffee to absorb excess carbon dioxide emitted by the coffee. In this manner, the roasted and ground coffee could be packaged almost immediately after grinding to minimize flavor volatilization. In conventional practice, roasted and ground coffee is "aged" in bulk to permit carbon dioxide produced in roasting to be expelled and thus avoid internal can pressurization. During this "aging," desirable coffee volatiles are lost. The use of the carbon dioxide scavenger was also supposed to permit the use of a lighter-gauge steel can, but this result was not necessarily achieved.

Calcium oxide, or lime, reacts with carbon dioxide. Packets of this compound are used extensively in containers for shipment of controlled-atmosphere fruit and vegetables. Calcium oxide has not been known to be incorporated into package materials. Companies such as Transfresh and Transicold, both truck and shipboard container operators, incorporate lime into packets in bunkers through which the internal atmosphere is circulated. This operation removes excess carbon dioxide produced by respiring fruit or vegetables and thus obviates the effects of excess carbon dioxide, including reduced pH and color and flavor changes.

CARBON DIOXIDE EMITTERS/GENERATORS

Sodium bicarbonate, when used together with citric acid and when wetted with water, generates carbon dioxide. This system was used in Europe during the late 1980s to generate carbon dioxide in modified-atmosphere packages of fish. "Gemella" was the name applied to the package used to incorporate the carbon dioxide emitter by its French paperboard carton converter developer, SPIC. The company is no longer involved in this product development.

During the 1980s, the concept of carbon dioxide emitters in package structures was intriguing and attracted considerable worldwide attention. The "Gemella" package was a tub/tray shape fabricated by thermoforming an inner liner into a printed paperboard shell. The thermoform contained a shelf on which the food rested. Beneath the shelf was a compartment in which was placed a dry, carbon dioxide–emitting chemical composed of citric acid and sodium bicarbonate. As the food product, usually a meat or fish, deteriorated biochemically, the protein molecules tended to contract, squeezing out aqueous purge that dripped through the open mesh shelf. The liquid reacted with the chemicals, releasing carbon dioxide at about the same rate at which atmospheric carbon dioxide was permeated out of the package. Thus the requisite preservative level of carbon dioxide within the package was maintained.

However proprietary this system was, almost all the packages were fabricated on single-unit up machinery that was fed manually. A much-publicized automatic erection machine never functioned as intended.

Results from the technical tests were quite good, with long-term refrigerated shelf life achieved. The economics of the package, with its "shelf" plus a compartment for chemicals, however, were not favorable for development. The package volume for the quantity of product contained was not welcomed by the consumer. The basic concept of generating carbon dioxide is, however, quite interesting from a technical perspective.

THE FUTURE

Other active packaging systems that might become available commercially are antioxidant-releasing films, flavor-absorbing and flavor-emitting systems, anti-fogging films, oxygen emitters, and light-blocking/-regulating compounds, all of which are currently being researched.

Further Reading

Ahvenainen, R., E. Hurme, K. Randall, and M. Eilame. 1995. "The Effect of Leakage on the Quality of Gas-Packed Foodstuffs with the Leak Detection " *VTT Research Notes 1683*. VTT Helsinki, Finland.

Anderson, H. S. 1989. "Controlled atmosphere package." U.S. Patent No. 4,842,875. June 27.

Andersen, H. J. and M. A. Rasmussen. 1992. "Interactive Packaging as Protection against Photodegradation of the Colour of Pasteurized Sliced Ham." *International Journal of Food Science Technology*, 27(1):1–8.

Anonymous. 1991. "Laminator for 'Active' Packaging." *Australian Packaging*, 39(7):16.

Anonymous. 1992. "Super-Sachet from EMCO." *Packaging Week*, 8(22):5.

Anonymous. 1993a. "Extending the Shelf Life." *Grocer*, 215(7094):46.

Anonymous. 1993b. "Absorbing Subject." *Packaging Week*, 9(16):18–19.

Anonymous. 1994. "Use of New Container Requiring No Deoxidizer." *Packaging Japan*, 19, July.

Appendini, P. and J. H. Hotchkiss. 1996. "Immobilization of Lysozyme on Synthetic Polymers for the Application to Food Packages." Presented at the annual meeting of Institute of Food Technologists, New Orleans, Louisiana, June 22–26.

Brody, A. L. and J. A. Budny. 1995. "Enzymes as Active Packaging Agents." In: M. L. Rooney (ed.) *Active Food Packaging*. Glasgow, UK: Blackie Academic & Professional, pp. 174–192.

Chiang, W. L., D. Chokshi, B. Tsai, and N. L. Venkateshwaran. 1996. World Patent Application W09640412; 1998. "Oxygen-scavenging Compositions and Articles." U.S. Patent 5,744,056. April 28.

Cicconci, J. P., E. S. DeCastro, and John B. Kerr. 1990. "Macrocyclic amine complexes for ligand extraction and generation." U.S. Patent 4,952,289, August 28.

Cook, John M. 1969. "Flexible film wrapper." U.S. Patent 3,429,717, February 25.

Courtland, Steven B., Gordon N. McGrew, and Lindell Richey. 1991. "Food packaging improvements." U.S. Patent 5,064,698, November 12.

Courtland, Steven B., Gordon N. McGrew, and Lindell Richey. 1992. "Food packaging improvements." U.S. Patent 5,126,174, June 30.

Fukazawa, R. 1980. "Methods of preventing spoilage of foods." Japanese Patent 23071180.

Hirata, Takashi. 1992. "Recent Development of Food Packaging in Japan." *The First Japan-Australia Workshop on Food Processing Proceedings*, Tsukuba Center, Tsukuba, Japan, February.

Hong, Kuo-Zong, Yong J. Kim, and W Cornell. 1994. "Barrier composition and articles made therefrom." U.S. Patent 5,281,360, January 25

Krochta, J. M. and C. De Mulder-Johnston 1997. "Edible and Biodegradable Polymer Films: Challenges and Opportunities." *Food Technology*, 51(2):61–74.

Naito, S. 1989. "Freshness Preservation of Processed Food." *Food Packaging* (Japan), 7:55–70.

Singh, A. and J. M. Schnur. 1994. "Polymerisable lipids for preparing vesicles that controllably release an encapsulant." U.S. Patent 5,366,881 A.

Appendix

TABLE A1. Some Active Packaging Material and Systems Suppliers.

Company and Address	Products
Alcohol Release Agents	
Bourbon Co , Ltd 5-61 Moritomo, Nishi-ku Kobe City, Hyogo Pref Japan	Encapsulated alcohol preservative powder that vaporizes; acts as antimicrobial agent in filter pouch paper; used with chocolate layer cake
Freund Industrial/Biddle Sawyer Corp 2 Penn Plaza New York, New York 10121	"Ethicap" microencapsulated food-grade ethyl alcohol; ethyl alcohol vaporization
Mitsubishi Gas Chemical Company, Inc 5-2, Marunouchi, 2-Chome Chiyoda-Ku, Tokyo 100 Japan	Synergistic effect of an oxygen scavenger, ethanol, and a natural preservative
Nitto Denko Corp 1-1-2 Shimohozumi, Ibaraki Osaka, 567-8680 Japan	Breathlon film, a multiporous polyethylene film for packaging preservatives such as agents that evolve ethanol
Nisshin Flour Milling 1-25, Kanda-Nishikicho Chiyoda-Ku, Tokyo 101-8441 Japan	Fry mix and bread crumbs with oxidizer and alcohol mixture
Showa Tansan Co , Ltd 3-23 Misaki-Cho 3-Chome, Chiyoda-Ku, Tokyo 101-0061 Japan	Ethanol spray

TABLE A1. (Continued)

Company and Address	Products
Aldehyde Scavengers	
Ciba Specialty Chemicals Corp 540 White Plains Road Tarrytown, New York 10541-9055	Ferrous iron-based oxygen scavengers for incorporation into plastics; additives for plastics used to store water
DuPont Polymers and Packaging Products Chestnut Run Plaza P O Box 80713 Wilmington, Delaware 19880	Aldehyde-absorbing additive in film extends shelf life of fatty foods
Hoffmann-LaRochè, Inc. 340 Kingsland St Nutley, New Jersey 07110-1199	Vitamin E antioxidant can reduce aldehydes
Tri-Seal/ZapatA Industries, Inc 900 Bradley Hill Road Blauvelt, New York 10913	Oxygen scavengers incorporated into closures for beer, etc , bottles
Antimicrobials	
Allied Resinous Products Inc Box 620 Conneaut, Ohio 44030	"Bacticlean" antibacterial additive combines several antimicrobial agents that destroy bacteria and fungi
Aso Co., Ltd.	Antibacterial polyethylene shrink film for drums
Daikoku Kasei Co 2-8-15 Nishi-Nakajima Yodogawa-ku, Osaka-shi, Osaka 532 Japan	"Piatech" film prevents mold and bacterial growth
EcoArt AG Felsenstrasse 4 CH-8808 Pfaffikon Switzerland	Antimicrobial agents
Ferro Corporation Ferro Chemicals Group Bedford Chemical Division 7050 Krick Road Walton Hills, Ohio 44146-4494	Micro-Chek antimicrobials
Givaudan-Roure Flavors 1199 Edison Drive Cincinnati, Ohio 45216	Giv-Gard DXN
Hitachi Borden Co 2-8-3 Kiba Koto-ku, Tokyo 135 Japan	Antibacterial food wrap: • "Antibacterial Hitachi Wrap" • "Antibacterial Wrap Food Guard"
Integrated Ingredients 10600 West Higgins Road, Suite 303 Rosemont, Illinois 60019-8748	NISAPLIN® controls bacterial spoilage

TABLE A1. (Continued)

Company and Address	Products
Kanebo, Ltd 20-20, Kaigan 3-Chome Minato-Ku, Tokyo 108-8080 Japan	"Bacte Killer"
Microban Products Co 11515 Vanstory Drive, Suite #110 Huntersville, North Carolina 28078	Microbicide additives for film; not FDA cleared
Mitsubishi Gas Chemical Company, Inc 5-2, Marunouchi, 2-Chome Chiyoda-Ku, Tokyo 100 Japan	Combination of oxygen scavenger, ethanol, and natural antibacterial
Mitsubishi International Corp Paper and Packaging Dept 520 Madison Ave New York, New York 10022	Antibacterial paper
National Taiwan University No 59, Lane 144 Keeling Rd , Sec 4 Taipei, Taiwan ROC	Antimicrobial edible film
Nichirei Co 6-19-20 Tsukiji Chuo-ku, Tokyo 104 Japan	Bactericide technology with Kobe Steel
Nimiko Co 4-19-9 Maebara-higashi Funabashi-shi, Chiba Z74 Japan	"Silvi Film" bacteriostatic
Nippon Soda Co , Ltd 2-1, Ohtemachi 2-Chome Chiyoda-Ku, Tokyo 100-8165 Japan	TiO_2 over a layer of pure silica is a surface treatment for glass that kills microorganisms when exposed to ultraviolet light It also breaks down odors
Okamoto Industries, Inc 3-27-12 Hongo Bunkyo-ku, Tokyo 113-91 Japan	Antibacterial food wrap: "Okamoto Super Wrap Antibacterial"
University of Alberta 114 St-89 Ave Edmonton, Alberta T6G 2M7 Canada	Investigation of potassium sorbate added to polyethylene film as antimicrobial agent
Sangi Group America 417 S Wall St #302 Los Angeles, California 90013	"Apacider" antimicrobial with silver

TABLE A1. (Continued)

Company and Address	Products
Sekisui Jushi Corp 4-4 Nishi-Tenma 2-Chome Kita-Ku, Osaka 530-0047 Japan	Wasabi antibacterial food-packaging material
Shin-Etsu Polymer Co , Ltd 4-3-5 Nihonbashi-honcho Chuo-ku, Tokyo 103 Japan	Antibacterial food wrap: "Antibacterial Polymer Wrap"
Shinanen New Ceramics Co , Ltd 4-22 Kaigan, 1 Chome Minato-ku, Tokyo Japan	Zeomic®
Mitsubishi International Corp Paper and Packaging Dept 520 Madison Ave New York, New York 10022	Zeomix; silver coated onto zeolite for incorporation into food contact layer
Mitsubishi Chemical Industries America, Inc 81 Main St White Plains, New York 10601	
Trico Industries, Inc P. O Box 806 Devils Foot Road Davisville, Rhode Island 02854-0806	Antimicrobial films
Tsuboi Co , Ltd Japan	Antibacterial carton with Wasa Power—PET film coated with natural wasabi made by Sekisui Chemical
Dr J S Paik Department of Food Science University of Delaware Newark, Delaware 19716	Method of imparting antimicrobial activity into packaging films for cheese and meat
Taiyo Chemical 2-16-10 Nihonbashi Chuo-ku Tokyo 103 Japan	Antimicrobial film
Try Company 43 Shige Numazu-Shi, Shizuoka 410-01 Japan	Antimicrobial film
Aroma Emitters	
Andrea Aromatics 150 Enterprise Ave Trenton, New Jersey 08638-0440	Concentrated compound of fragrance additives from extraction of natural leather and fortified with economical ingredients to mask "off odors " It also has antioxidant properties

TABLE A1. (Continued)

Company and Address	Products
Dragoco 10 Gordon Drive Totowa, New Jersey 07512	
Techno International 130 West 37th St New York, New York	Fretek ethanol-emitting antimicrobial plus aroma emitter
Japan Liquid Crystal Co 2-10-3 Honkomagome, Bunkyo-ku Tokyo 113 Japan	
M A T Co , Ltd 2-11-4 Tamagawadai, Setagaya-ku Tokyo 158 Japan	
Okamoto Industries 3-27-12 Hongo Bunkyo-Ku, Tokyo 113-91 Japan	
Buffers	
Feinkost Ingredient Company, Inc 103 Billman St Lodi, Ohio 44254	Mayonat® VM™ buffer claimed to inhibit yeast, mold, and bacterial growth
Carbon Dioxide Absorbers or Emitters	
Mitsubishi Gas Chemical Company, Inc 5-2, Marunouchi, 2-Chome Chiyoda-Ku, Tokyo 100 Japan	• Ageless®—reacts with carbon dioxide (contains calcium hydroxide) for fresh-roasted coffee • Fresh Lock sachet has both oxygen and carbon dioxide scavenger
Nippon Unican Co. Asaha Tokai Bldg Otemachi, Chiyoda-ku Tokyo 100 Japan	Polyethylene compounded with a deodorant of green tea
United Desiccants—Gates P O Box 32370 Louisville, Kentucky 40232	CO_2SORB™ CO_2-adsorbing packets
SPIC S A Mallasherbes 41, Av De L'Agent Sarre 92 700 Colombes France	Verifrais system emits carbon dioxide; based on sodium bicarbonate and citric acid

TABLE A1. (Continued)

Company and Address	Products
Deodorants/Deodorizers	
Good Life Japan	Yakimono Meijin; eliminates unpleasant odors as well as absorbing fats and grease from grills (meat and fish)
Nippon Soda Co , Ltd 2-1, Ohtemachi, 2-Chome Chiyoda-ku, Tokyo 100-8165 Japan	Surface treatment for glass uses TiO_2; when exposed to UV light, destroys microorganisms and also breaks down odors
UOP Molecular Sieves Old Saw Mill River Road Rt 100C Tarrytown, New York 10591	Anti-yeast, ABSCENTS®, SMELLRITE deodorizing powders
Masadaya KK Japan	Binchotan; Ofuroyo Binchotan (charcoal from oak trees grown in Wakayawa Prefecture), acts as a deodorizer and dehumidifier
Narui Norin KK Fukushima Prefecture Japan	Osumi Monogataxi
Johns Hopkins University 624 N Broadway Baltimore, Maryland 21205	Odorant binding protein (OBP)
Structural Biology Laboratory National Center for Scientific Research Marseilles, France	Odorant binding protein
Toagosei Co , Ltd 14-1, Nishi-Shinbashi, 1-Chome Minato-Ku, Tokyo 105-8419 Japan	Novaron; inorganic silver series of antibactarium agents that result in molded products within smells
Lintec Corp 23-23, Hon-Cho Itabashi-Ku, Tokyo 173-0001 Japan	Paper label with deodorant function; deodorant manufactured by Minato Sangyo
Morishita Jintan Co 1-30, Tamatsukuri, 1-Chome Chuo-Ku, Osaka 540-8566 Japan	
Ginpo Pack KK 1 Kanda-Hira Hawacho Chiyoda-ku, Tokyo Japan	Polyethylene pack with magnesium oxide to keep foods fresh and odor free
Mitsubishi Gas Chemical Co , Inc 5-2, Marunouchi, 2-Chome Chiyoda-Ku, Tokyo 100 Japan	Odor-absorbing film containing polyphenol derived from tea leaves

TABLE A1. (Continued)

Company and Address	Products
	Combination of synergistic effect of an oxygen scavenger, ethanol, and a natural preservative; prevents growth of aerobes; no off-odors produced in the packaged food
Naarden International USA 919 Third Ave. New York, New York 10022	Polyvinyl chloride; polymer flavor/ fragrance concentrates; carrier systems for high loads of flavors or fragrances; uses a proprietary polymer alloy technology
International Paper Company Two Manhattanville Road Purchase, New York 10577	Paper that traps odor; uses Abscents® powder from UOP

Desiccants

Capitol Specialty Plastics, Inc 2039 McMillan Street Auburn, Alabama 36830	Removes moisture using desiccant-loaded plastic in vials, bottles, trays, and strips
Grace Davison/W R Grace & Co 5500 Chemical Road Baltimore, Maryland 21226-1698	Flo-Tech™ hot-melt desiccant
Hammond W. A Drierite Co , Ltd P. O Box 460 Xenia, Ohio 45385	Drierite desiccants from calcium sulfate
Multisorb Technologies, Inc 325 Harlem Rd Buffalo, New York 14224-1893	Removes moisture, also has oxygen scavengers FreshPax™, FreshMax®, DesiMax®, MiniPax®, Sorbicap™
United Desiccants—Gates (now Süd-Chemie Performance Packaging) P. O Box 32370 Louisville, Kentucky 40232	Removes moisture

Ethylene Absorbers (Scavengers)

Dry Company Ltd. Japan	Cool Chill packs of nylon/polyethylene film; nylon is impregnated with a protein extracted from Japanese horseradish, which also suppresses the production of ethylene gas
Ethylene Control, Inc 8252 E Oinubu P O Box 571 Selma, California 93662	For fruits/vegetables; sachets of potassium permanganate; working in cooperation with University of California–Davis

TABLE A1. (Continued)

Company and Address	Products
Evert-Fresh Corp 3701 Kirby Dr #728 Houston, Texas 77098-0390	Low density polyethylene film impregnated with an ethylene gas-absorbing mineral called oxa, a cousin to zeolite; tested in Philippines by Dole Packaged Foods Co for asparagus
C Itoh & Co One Cascade Plaza, Room 1904 Akron, Ohio 44308	Orega Pack, Orega Films, polyethylene film line with ethylene-absorbing filler for fruits/vegetables; filler includes pumice, zeolite, activated carbon and metallic oxides; developed and being used in Korea
Cho Won Wha Sung Co , Ltd Cho Yang Bldg 58-4HO, Samsung-Dong, Kangnam-Ku, Seoul Korea	Ever-Fresh Bag; allows ethylene to pass through the bag
Cho Yang Heung Son Co , Ltd Cho Yang Bldg 58-4HO, Samsung-Dong, Kangnam-Ku, Seoul Korea	
Cal Agri Company Air, Repair Products, Inc P O Box 10067 Stafford, Texas 77477	"Air-Repair" ethylene gas scrubbers
Food Science Australia Delhi Rd , P O Box 52 North Ryde, NSW 2113 Australia	For fresh fruit/vegetables plus fungicide
Sueo Urushizaki National Institute of Agrobiological Resources Isukuba Science City Ibaraki 305 Japan	
Nippon Unican Co Asaha Toka Bldg. Otemachi, Chiyoda-ku Tokyo 100 Japan	Compounded special polyethylene bags for food packaging; ceramic incorporates that absorb ethylene; another is compounded with a deodorant of green tea
Nissho Corporation 9-3, Honjo-Nishi, 3-Chome Kita-Ku, Osaka 531-8510 agents	Supplier of film and bags; Hosenka Pack, which incorporates ethylene absorptive and for infrared radioactive ceramics and anti-haze agents

TABLE A1. (Continued)

Company and Address	Products
Nitto Funka Trading Co , Ltd Japan	Zeopet No 40 N; masterbatch for producing freshness-keeping deodorant film, a compound of polyethylene and activated zeolite powder; can absorb ethylene and carbon dioxide
Rengo Packaging Systems Co , Ltd Sumito Shoji Mitoshiro Bldg , 1-Kanda Tokyo T101 Japan	Green Pack—sachet with potassium permanganate embedded in silicon oxide
Warenhandels GmbH 1130 Vienna Hietzinger Hauptstrasse 50 Austria	ProFresh fresh-keeping and malodor control masterbatch to polyethylene for fruits, vegetables, flowers, meat/fish, and trash bags

Fruits and Vegetables

Ikulon Corporation	Ikilun-1 Natural Mineral, fresh vegetable activating mineral
Fuso Corporation Japan	Ikilun-1 and polyethylene film containing ceramics (for infrared emitters and ethylene absorbers)
Shiba & Co , (N Z) Ltd Auckland, New Zealand	

Fungicides

Food Science Australia Delhi Road, P O Box 52 North Ryde, NSW 2113 Australia	Film with fungicide for fresh fruit
Horticultural Research and Development Corp Sydney, Australia	
Sekisui Jushi Corp 2-4-4 Nishi Tenma, Kita-ku Osaka City, Osaka Prefecture Japan	Wasabi

Germicides

Gunze Sangyo Inc Kyoto Br Muromachi-dori Oike-sagaru Nakagyo-ku Kyoto City, Kyoto Prefecture Japan	Germicidal film to prevent bacterial growth

TABLE A1. (Continued)

Company and Address	Products
Mold and Yeast Inhibitors	
Bourbon Co , Ltd 5-61 Moritomo, Nishi-ku Kobe City, Hyogo Pref Japan	Anti-mold; preservative in a pouch to prevent mold and bacteria in packaged layer cake; encapsulated alcohol powder; antimicrobial agent
Feinkost Ingredient Company, Inc 103 Billman St Lodi, Ohio 44254	Anti-yeast; Mayonat® VM™
Mitsubishi Gas Chemical 5-2, Marunouchi, 2-Chome Chiyoda-Ku, Tokyo 100 Tokyo, Japan	Anti-yeast, synergistic effect of oxygen scavenger, ethanol, and natural preservative
Multisorb Technologies, Inc 325 Harlem Rd Buffalo, New York 14224-1893	Anti-yeast; SorbiCAP™ sorbent canister for pharmaceutical and diagnostic product packaging
Nimiko Co 4-19-9 Maebara-higashi Funabashi-shi, Chiba 274 Japan	Anti-mold; "Silvi Film" for antimold and antibacterial applications
Shinanen New Ceramics Co Japan	Anti-mold; Zeomic®; restrains growth of bacteria mold or yeast Anti-yeast; Zeomic®
Watson Foods Co , Inc West Haven, Connecticut	Anti-mold; "No Mold" mold inhibitor
Oxygen Scavengers	
Alcoa Closure Systems International Inc 1604 E Elmore St Crawfordsville, Indiana 47933	ALCOA CSI—oxygen displacer closure system for beer applications
BP Amoco P L C Mail Code C-T 150 West Warrenville Road Naperville, Illinois 60563	Amosorb® 3000 organic oxygen absorber for polyester
Chevron Chemical Company LLC Petrochemicals & Plastics Division P O Box 3766 Houston, Texas 77253-3766	Polymeric oxygen-scavenging system
Richmond Technology Center 100 Chevron Way Richmond, California 94802-0627	
Ciba Specialty Chemicals Corp Additives 540 White Plains Road P O Box 2005 Tarrytown, New York 10591-9005	Ciba™ Shelfplus™ oxygen inorganic-based oxygen absorbers for polyethylene and polypropylene

TABLE A1. (**Continued**)

Company and Address	Products
Crown Cork & Seal Co , Inc One Crown Way Philadelphia, Pennsylvania 19154-4599	Starshield™ oxygen scavenger PET bottle technologies
Cryovac Sealed Air Corporation Rogers Bridge Road, Bldg A P O Box 464 Duncan, South Carolina 29334-0464	Polymeric scavenging system
Desiccare, Inc 10600 Shoemaker Avenue, Bldg C Santa Fe Springs, California 90670	O-Buster® sachets imported from Taiwan; ferrous iron based
Food Science Australia (CSIRO)/Southcorp Packaging P O Box 52 North Ryde, NSW 1670 Australia	ZerO$_2$™ oxygen-scavenging plastic package materials
Mitsubishi Gas Chemical America, Inc 520 Madison Avenue, 25th Floor New York, New York 10022	Ageless® iron-based sachets and labels
Mitsubishi Gas Chemical Company, Inc Mitsubishi Building 5-2 Marunouchi 2-chome Chiyoda-ku Tokyo Japan	Ageless® iron-based sachets and labels
Multisorb Technologies, Inc 325 Harlem Road Buffalo, New York 14224-1893	Resin-bonded oxygen absorbers, FreshPax™ and FreshMax® iron-based sachets and labels
Standa Industries F 14050 Caen Cedex France	ATCO® iron-based sachets and labels
Toyo Seikan Kaisha, Ltd 3-1 Uchisaiwaicho 1-Chome Chiyoda-Ku Tokyo 100, Japan	Oxyguard iron-based film
Tri-Seal International, Inc A Tekni-Plex Company 900 Bradley Hill Road Blauvelt, New York 10913	Tri-So$_2$RB® closure liner materials for food and beverages; Dar Eval™ EVOH-based barrier vesin; Dar Extend for closures and crowns
W R Grace & Co -Conn. Grace Performance Chemicals-Darex 55 Hayden Avenue Lexington, Massachusetts 02421-7999	OST™ liner compound for beer and beverages; Dar Eval™ EVOH-based barrier resin; Dar Extend for closures and crowns

TABLE A1. (Continued)

Company and Address	Products
ZapatA Industries, Inc 4400 Don Cayo Drive Muskogee, Oklahoma 74403	TS-ASR9 PureSeal® Compound

Preservatives

Company and Address	Products
Minato Sangyo Co Ltd Institute of Chemical Technology of the Board of Industrial Technology Japan	"Anico-S" freshness preservative
Mauri Laboratories 10600 West Higgins Rd , Ste. 103 Rosemount, Illinois 60018	Nisaplin™ preservative for beer, wine, and food

Mitsubishi Electric Corporation
2-3, Marunouchi, 2-Chome
Chiyoda-Ku, Tokyo 100-8310
Japan
	Anions (mainly O_2) to prevent
	    multiplication of microorganisms that
	    decay foodstuffs

Argus Chemical 1500 E Lake Cook Rd Buffalo Grove, Illinois 60089	Myacide BT preservative

Nitto Denko Corp
31 Mori Bldg , 5-7-2 Kojimachi
Chiyoda-ku, Tokyo 102
Japan
	Breathlon film; multi-porous
	    polyethylene film makes it useful for
	    packaging of preservation such as
	    agents which evolve ethanol

Dai Nippon Printing Co
1-14, Ichigaya-Kagacho
Shinjuru-ku, Tokyo 162
Japan
	"Antibiotic film"; has a metal-bonded
	    zeolite on inside surface; ozone or
	    active oxygen produced as it comes
	    through the film, which in turn kills
	    surface microbes

Viskase Corporation 6855 W 65th St Chicago, Illinois 60638	Sausage casing for large bologna; casing contains sorbate and glycol

Water Vapor Control

Company and Address	Products

Grace Davison
5500 Chemical Road
Baltimore, Maryland 21226-1698
	CondensationGard™ silica gel
	    desiccant that adsorbs free water
	    without desiccating the food

Humidi-Pak 865 Parkview Ave St Paul, Minnesota 55117	Controls relative humidity to 70% for cigar preservations

Index

Milton Keynes UK
Ingram Content Group UK Ltd.
UKHW020028071024
449327UK00032B/2972